Ice Age Extinction

Ice Age Extinction

Cause and Human Consequences

Jim Snook

Algora Publishing
New York

© 2008 by Algora Publishing.
All Rights Reserved
www.algora.com

No portion of this book (beyond what is permitted by
Sections 107 or 108 of the United States Copyright Act of 1976)
may be reproduced by any process, stored in a retrieval system,
or transmitted in any form, or by any means, without the
express written permission of the publisher.
ISBN-13: 978-0-87586-557-7 (trade paper)
ISBN-13: 978-0-87586-558-4 (hard cover)
ISBN-13: 978-0-87586-559-1(ebook)

Library of Congress Cataloging-in-Publication Data —

Snook, Jim, 1933-
  Ice age extinction : cause and human consequences / Jim Snook.
    p. cm.
  Includes bibliographical references and index.
   ISBN 978-0-87586-558-4 (hard cover: alk. paper) — ISBN 978-0-87586-557-7 (trade paper: alk. paper) — ISBN 978-0-87586-559-1 (ebook) 1. Extinction (Biology) 2. Glacial epoch. 3. Climatic changes—Environmental aspects. 4. Atmospheric carbon dioxide—Physiological effect. I. Title.

QE721.2.E97S66 2007
576.8'4—dc22
                           2007016276

Printed in the United States

Acknowledgements

I would like to thank my daughter, Lorrie Snook, who was my native guide through the literary jungle; Barbara Brown, who gave me early direction; and Jeline Harclerode, who put the book into publishing form.

Maps and Diagrams

Figure 1. Generalized Vegetation Formations of North America before Glacial Breakup
Figure 2. Loess and Sand Dune Deposits in North America Formed Near the Time of Glacial Breakup
Figure 3. Generalized Vegetation Formations of North America Today
Figure 4. Global Distribution of Loess and Sand Dunes
Figure 5. Idealized 100,000 Year Glacial Diagram
Figure 6. Full Data Diagram of Idealized 100,000 Year Glacial Cycle
Figure 7. Diagram of Centrifugal Force
Figure 8. Carbon Dioxide and Population Increases 1800-2000

Table of Contents

Preface	1
Chapter 1. Introduction	3
The Last Extinction Compared to Other Times of Extinction	4
How the Last Extinction Relates to the Ice Age	5
Life Changes near the End of the Last Ice Age	7
Background Data	8
Low Atmospheric Carbon Dioxide Effect	9
Looking Ahead	10
Chapter 2. Transitions Relating to Extinction	11
Exposed Land	11
Greatest Glacial Extent	12
Changing Plant Distribution and Character	14
Woolly Mammoth and Relatives	17
Large Animal Distribution	19
Chapter 3. Extinction	21
Melting of the Continental Glaciers	21
Large Animal Extinction	23
Loess and Sand Dune Deposits	27
Dust and Sand Storm Effects on Animals and Humans	29
Chapter 4. How the Glacial Cycle Works	31
Heat Cycles	32
Energy for the Glacial Cycle	33
How Heat is Utilized on Earth	36
How a Glacier Works	38
Types of Glaciers	38
Idealized 100,000 Year Glacial Cycle	41

Interglacial Period	42
Accumulation Phase	43
Background Information for Full Data Diagram	44
Dormant Phase	45
Active Phase	46
Glacial Breakup Phase	46
CHAPTER 5. OCEAN CHANGES RELATING TO GLACIATION AND EXTINCTION	49
Atmospheric Gases in the Oceans	50
Prelude to Extinction in the Active Phase	53
Changes to the Oceans during Glacial Breakup	54
Meltwater Distribution during Glacial Melting	55
Carbon Dioxide Absorbed by the Oceans	56
After the Glacial Breakup	57
CHAPTER 6. CHANGES IN THE ATMOSPHERE DURING THE LAST EXTINCTION	59
Origin of the Atmosphere	59
Evolution of the Atmosphere	61
Shape of the Atmosphere	63
Tree Lines	65
Late Glacial Cycle Atmospheric Changes Leading to Extinction	67
The Atmosphere during the Interglacial Period	68
Particles in the Atmosphere	69
CHAPTER 7. GLACIAL CHANGES TO THE LAND AFFECTING LIFE	71
What Glaciers Do to the Land	71
Land Changes from Glacial Breakup and Melting	73
Land Changes Left Behind After Glaciation	74
Events Leading to Loess and Sand Dune Deposits	75
Origin of Loess and Sand Dune Deposits	76
Land Changes from Lowering Sea Level	78
Land Changes from Rising Sea Level	79
CHAPTER 8. CHANGES IN PLANTS LEADING TO EXTINCTION	81
Plant Changes throughout Geological History	82
Plant Changes in the Last Glacial Cycle	83
Plant Changes Leading to Extinction	85
Some Plant Life Changes Associated with Extinction	88
Plant Changes since the End of the Last Ice Age	89
CHAPTER 9. LARGE ANIMAL EXTINCTION ASSOCIATED WITH GLACIAL MELTING	91
Origin and Evolution of Animals	91
Animal Life in the Pleistocene	94
Animal Life in the Last Glacial Breakup Phase	96
Extent of Extinction	97

Key Points for Extinction	99
Animal Life after the Extinction	102
CHAPTER 10. THE HUMAN CONDITION DURING EXTINCTION	103
Origin and Development of Early Humans	104
Humans in the Pleistocene	104
Humans in the Active Phase Prior to Extinction	107
Humans in the Last Glacial Breakup	108
CHAPTER 11. FIRST BIG INCREASE IN ATMOSPHERIC CARBON DIOXIDE AND POPULATION	111
Humans in the Current Interglacial Period	112
Rapid Increase in Atmospheric Carbon Dioxide and Human, Plant, and Animal Population	113
Agriculture	114
Irrigation and Nutrients	115
Growth of Civilization	117
Warm Spell before the Little Ice Age	118
Little Ice Age	118
Human Response to the Little Ice Age	121
CHAPTER 12. CURRENT BIG INCREASE IN ATMOSPHERIC CARBON DIOXIDE AND POPULATION	125
Large Increase in Atmospheric Carbon Dioxide and Population in the Last 150 Years	125
Crop Yield Increase Caused by Increase in Atmospheric Carbon Dioxide	128
Wheat Crop Sketches: 1918–1998	129
(Source: Kansas Agriculture Statistics Service)	*129*
Other Factors Associated with Food Supply	131
Factors Altering Human Population Growth	134
The Oceans Rising or Seashores Sinking	136
Plant and Animal Changes with Increased Atmospheric Carbon Dioxide	138
Attitude Changes with Increased Atmospheric Carbon Dioxide and Population	140
CHAPTER 13. THE WAY WE ARE AND WHERE WE ARE HEADING	143
Where We Are Today	143
Human Population	*144*
Energy	*146*
Oil	*146*
Tar Sands	*148*
Oil Shale	*149*
Natural Gas	*149*
Methane Clathrates	*151*
Coal	*152*
Nuclear	*153*
Hydroelectric	*153*
Geothermal	*154*
Alcohol, Wood, and Waste	*154*

Wind	*155*
Solar	*155*
Carbon Dioxide	*156*
Land and Water	*157*
Food	*159*
Where We are Heading	159
Plants and Animals Going Forward	*161*
Historic Data and Future Projections	161
Sometimes People Can be an Arrogant Lot	163
CHAPTER 14. THE FUTURE	165
Life in the Age of Declining Fossil Fuels	165
Life in a Time of Declining and Aging Population	167
Life After the End of the Fossil Fuels	168
Life in the Age of Declining Atmospheric Carbon Dioxide	170
Life in the Coming Glacial Cycle	171
Conclusion	172
SOURCES	175
INDEX OF NAMES	179
INDEX OF SUBJECTS	185

PREFACE

The writing of this book was more accidental than planned. After retirement as a petroleum geologist, there was time to study a broad range of subjects. These included the earth and its volcanoes, earthquakes, glaciers, oceans, atmosphere, and how these affected all life. Along this journey of discovery, the cause of the large animal extinction near the end of the last ice age was identified. This led to focusing on the mechanisms of life and extinction. Understanding some of these mechanisms came from past personal experiences.

My purpose in writing is threefold: First, to explain the cause of the last extinction and what this means for plants, animals, and humans; second, to demonstrate that personal observation and experiences are important tools. In many cases what is seen, discovered, and understood is far more valid than media interpretations. The third purpose is to present information in an interesting reader-friendly way with explanations and examples. Science should never be intimidating. Science is all around us and all we have to do is look. Plants, animals, humans, and the earth are all fascinating to study.

The sources used are basic science including the chemical and physical properties of water, air, and earth; existing literature; and personal observation and experiences.

Scientific knowledge moves forward by discovery and innovative thinking, not by consensus building. Anytime there is a challenge to the current scientific thinking, even when accepted, it is not an endpoint of scientific research. It is a starting point for research in a different direction. Only time will tell if this is a starting point for future research by the next generation.

Chapter 1. Introduction

We are living on a planet that provides all the needs for abundant life. The earth is extremely diverse with myriad environments from tropical rain forests to glacial areas, and from high mountains to deep oceans. A profusion of plants and animals of all sizes, shapes, and forms fill most ecological niches.

However, life on earth comes with no guarantees. Daily life for plants, animals, and humans can be harsh with "survival of the fittest" being a law of nature. All life is at risk from the extremes of weather, changing climate, and natural forces such as earthquakes, tsunamis, and volcanoes. Animals eat plants and other animals and humans eat both. Life can be dangerous, and sometimes whole populations of plants and animals are at risk.

Throughout geological history, there have been major events which have killed off not only individual plants and animals, but whole species and genera. These events, referred to as extinction, are usually very complex. Many people believe that we are now living in a time of extinction, and this, by some definitions, is true. However, the last extinction occurred near the end of the last ice age between 26,000 and 9,000 years ago. This extinction was when most of the large land animal genera of the world disappeared. In North America this included the woolly mammoth, saber tooth cat, mastodon, North American horse, North American camel, and twenty-nine other large animal genera. Since there are many species in one genus, we talk of "genera loss" for simplicity. In addition to the loss of whole genera, there was a loss of some species while other species in the genera survived. The cause of the extinction of the woolly mammoth and other large animals of the time has always been for me a nagging question. As I researched the extinction, it turned out to be much worse than I had imagined.

Before the last extinction, the animal world was much different from the way it is today. There was a far greater variety of strange and unique large animals. There were four times as many large animal genera in North America as there are today; in South America there were five times as many large animal genera as today; in Australia, there were seven times as many; in Europe and Asia there were twice as many; even in Africa there were 14% more large animal genera as there are today.

The Last Extinction Compared to Other Times of Extinction

Before the last extinction there had been five major and many minor episodes of extinction in the preceding 500,000,000 years. Many genera in both the oceans and on land were killed off in these episodes. For example, in each major extinction 40% to 80% of the animal genera in the oceans became extinct. Most of these episodes also wiped out many land animals, including the extinction at the end of the Cretaceous period when the dinosaurs disappeared. Some of these times of extinction were rapid, but some took millions of years. Some coincided with great volcanic eruptions or mountain building. Other times of extinction were accompanied by sharp changes in sea level and climate. Some were probably associated with asteroid or comet impacts.

However, the extinction that occurred near the end of the last ice age was very strange and much different from any previous extinction. There was a great deal of glacial melting and ice deposition occurring simultaneously with the start of the last extinction. Under these conditions one would think that life for the large animals would start to improve. Instead, it became lethal. The large animal extinction occurred only on land and did not affect ocean animals. Most large animal genera were wiped out on all continents except Africa where there were smaller losses.

The extinction of large animals was preceded by a reduction in size in many genera. This would suggest that the animals' food supply diminished. For the most part, the animals that became extinct were those that grazed on grass, browsed on shrubs and trees, or were carnivores that ate the animals that grazed and browsed. Most of the very small animal genera survived the extinction, indicating that there may have been enough food in some areas for a few small animals to eat. These factors suggest that the large animal extinction was associated with melting of the glaciers and a loss of food supply.

Chapter 1. Introduction

How the Last Extinction Relates to the Ice Age

Over the last 200 years, people have been studying glaciers and glacial deposits throughout the world. Modern glaciology got its start in Europe in the early nineteenth century. A Swiss naturalist, Louis Agassiz, is known as the father of glaciology. Although he did not come up with the theory that glaciers had been more extensive, his work in Switzerland, other parts of Europe, and the United States brought acceptance of the theory. Scientists of that time found numerous glacial features such as drumlins or erratic boulders out in front of the active mountain glaciers out in front of the active mountain glaciers. They tracked these features back to the glaciers and could see many of these features forming in the glaciers. With this knowledge, they could recognize glacial features in areas where the glaciers had melted away long ago.

In two centuries of studying glaciers, geologists have determined that global glaciation occurs in cycles composed of a glacial period and an interglacial period. During the last glacial period, glacial ice covered more than a quarter of the land surface of the earth. In the interglacial periods much of the glacial ice retreats, that is, melts away, leaving only the permanent continental glaciers of Antarctica and Greenland, along with some smaller glaciers that come and go. The interglacial periods are the exception, not the rule. During the last 2,000,000 years, the earth has been in glacial periods most of the time. However, interglacial periods are the most advantageous time for life to flourish. We are extremely fortunate to be living during an interglacial period.

There were a number of glacial cycles during the Pleistocene Epoch, which covered most of the last 2,000,000 years. Some glaciation has also occurred at various times in the last 400,000,000 years. Each cycle of Pleistocene glaciation took a somewhat different path and covered different sized areas. During some glacial cycles, glaciers advanced onto land that had no previous glaciation. However, the glaciation of the Pleistocene is believed to have covered more land area with greater volumes of ice and lasted longer than any previous glaciation. Today we are believed to be about two-thirds of the way through the current interglacial period.

The last glacial cycle took about 100,000 years to complete. The climate 100,000 years ago was similar to what it is today. This was followed by an ice advance which covered most of the northern portions of North America along with large areas of northern Europe and Russia. Mountain glaciation was extensive in many part of the world, including most of the higher Rocky Mountains. There was also significant glaciation in the Southern Hemisphere covering much of Chile and western Argentina. The glacial mass of Antarctica was significantly larger than it is today. The end of the glacial period occurred when most of the continental glaciers broke up and melted.

About 28,000 years ago we received more energy from the sun and the atmosphere warmed. The surfaces of the oceans in the equatorial and mid latitudes were also warming. There was considerable melting of the glaciers, causing a slow rise in sea level. Abundant snowfall was also building glaciers. Glacial expansion occurred in areas such as the southern side of the North American ice sheets where glacier building exceeded melting. Glacial contraction from melting with little glacier building occurred in Antarctica. In the high latitudes, melting of the glaciers put cold meltwater on the surface of the oceans. With such cold surface temperatures, water didn't evaporate so there was little moisture in the atmosphere and not much snow falling on the glaciers near the poles. With the surface of the oceans warming in the low and mid latitudes, there was a great deal of evaporation and increasing moisture in the air. This caused much more snowfall over the glaciers farther away from the poles. With glacier building exceeding melting, they grew and expanded. This caused somewhat of a paradox with polar glaciers contracting and mid latitude glaciers expanding.

By 18,000 years ago, ice dissipation far exceeded ice formation and all glaciers rapidly started to recede. On average, about 1% of the ice volume melted every sixty years. Where the glaciers flowed to the oceans, large icebergs broke off. This is referred to as "calving icebergs." As the oceans rose, ice sheets floated off the ocean floors, much like a stranded boat is floated off the bottom by a rising tide. Then these ice sheets floated out to the open oceans, where they melted. By about 12,000 years ago most of the glaciers outside of Antarctica and Greenland had melted and the oceans had risen about 350 feet. That point 12,000 years ago was the transition from the glacial period to the interglacial period.

The large animal extinction is believed to have started in Australia about 26,000 years ago, continuing off and on until about 15,000 years ago. This extinction happened as major melting was occurring on the large ice sheet of East Antarctica, which is south of Australia. In North America, South America, Europe, Asia, and Africa, the large animal extinction occurred near the end of the glacial melting about 16,000 to 10,000 years ago. Changes in the oceans and atmosphere near the end of the last ice age are intimately associated with the large animal extinction.

The glacial cycle is not just an ice thing; it is related to everything else on Earth. The forces that cause the glacial cycle also create basic changes in the oceans, atmosphere, land, plants, animals and humans. There are several geological puzzles associated with the glacial cycle. One of these puzzles is what causes the glacial cycle and how does it work.

Chapter 1. Introduction

The driving force of the glacial cycle is energy, with the sun providing almost all of the energy. When the amount of energy reaching the atmosphere, oceans and land changes, then other exciting changes happen throughout the world. We will investigate how changes in the heat energy input affects the internal dynamics of the oceans and the atmosphere. We will see what changes cause massive amounts of water to be evaporated from the oceans which falls as precipitation and forms the glaciers that pile up on the continents. The glacial advances and retreats have a major impact on the land surface. When the oceans are lowered, erosion cuts deep valleys throughout the land surface and as the oceans rise these valleys are filled with sediment. We will take a closer look at the uncovering and drowning of the continental margins. We will solve the puzzle of what caused the large windblown dust deposits called loess and sand dune fields that were formed at the end of the last ice age. The wind blown dust deposits provide the fertile farmland creating the bread basket for many countries.

LIFE CHANGES NEAR THE END OF THE LAST ICE AGE

There are three life changes associated with the end of the last ice age that we will investigate. The first life change is why so many large animal genera became extinct at the end of the last ice age. No evidence has been found of mass extinction at the end of the previous ice ages, so climate alone is not the answer.

The second is why there is such plant diversity today in the tropical rain forests, close to the equator, and lack of plant diversity in temperate rain forests in other parts of the world. There are more than ten times as many plant species in a square mile of tropical rain forest as there are in a square mile of temperate rain forest. Plant diversity on land is greatest within 20 degrees of the equator and relatively close to sea level. The plant diversity diminishes as you move away from these areas toward the higher latitudes, and it becomes very sparse in the high latitudes.

The third life change is that most of the people disappeared during the later stages of the ice age. Also, there was an explosion of human population and migration after the end of the last ice age. Fossil and artifact evidence shows that people lived in many places throughout Europe, Asia, and Australia during much of the last glacial period. Examples of artifact evidence include stone and bone tools, stone-cut bones, hearths, and even art. There is some artifact evidence that suggests people lived in the western hemisphere thousands of years before the end of the last ice age. However, the evidence diminishes for people living in many of these areas in the later half of the glacial breakup phase of the last ice age. At about the same time the large animals were becoming extinct, humans disappeared from the same areas. However, humans started returning to these areas near the end of the large animal extinction. There appears to have been a rapid expansion of population, and large migrations of people re-populating the continents starting about 11,000 years ago.

To identify the cause of these life changes, we will look at some background data leading to the large animal extinction.

Background Data

In studying how plants, animals, and humans related to the glacial cycle, it became evident that there were complex things going on that we did not understand. A clue to some of these mechanisms comes from the study of carbon dioxide.

Life functions of plants are complex requiring sunlight, water, nutrients, and carbon dioxide. The net effect is that plants need carbon dioxide to live in much the same way that animals need oxygen. Plants take carbon dioxide from the air and release oxygen. Animals take oxygen from the air and release carbon dioxide. Without sufficient carbon dioxide available in the atmosphere for plant use, plants will die. Each plant species has a minimum atmospheric carbon dioxide requirement below which it cannot survive. For about 3,500,000,000 years, plants have been taking in carbon dioxide and expelling oxygen. At the time that plants first appeared in the oceans, the atmosphere was about 70% carbon dioxide or 700,000 parts per million. There was essentially no oxygen in the atmosphere. Because plants have been taking carbon dioxide out of the atmosphere and oceans for about 3,500,000,000 years, the carbon dioxide content of the atmosphere is down to about 360 parts per million. Oxygen, on the other hand, has increased to about 21% or 210,000 parts per million atmospheric concentration over the last 3,500,000,000 years. Essentially, all the oxygen in the atmosphere has come from plants.

Carbon dioxide is the most mobile and variable gas in the atmosphere and oceans. It is very soluble in water and is far more soluble than any other major gas in the atmosphere. Carbon dioxide is seventy-three times more soluble in cold water than nitrogen and thirty-five times more soluble in cold water than oxygen. Proportionately more carbon dioxide than any other component of the atmosphere is absorbed by the oceans' surface and rainwater, which ends up in the oceans. There is sixty times as much carbon dioxide in the oceans as there is in the atmosphere.

Chapter 1. Introduction

Carbon dioxide is twice as soluble in cold water as it is in warm water. When the surface of the oceans are the coldest, then carbon dioxide will move from the atmosphere to the oceans. When this happens, there is less carbon dioxide in the atmosphere to support plant life at the higher elevations and higher latitudes. When the surfaces of the oceans are the warmest, carbon dioxide is released from the oceans to the atmosphere, and plant life can flourish over more areas of the earth.

In addition to carbon dioxide concentrations in the atmosphere, we also need to look at where most of the air is located. The atmosphere is about twice as thick over the equator as it is over the poles. This distortion is caused by centrifugal force from the rotation of the earth. The thinner air has less carbon dioxide available for plant growth closer to the poles. The thicker air near the equator provides carbon dioxide to the plants for more abundant growth.

Low Atmospheric Carbon Dioxide Effect

The most destructive force of nature on plants and animals at the end of the Pleistocene was something so apparently benign that if it occurred today, it probably would go unnoticed by the average human until our food source started to disappear. This disaster was a gradual drop in atmospheric carbon dioxide to levels that could not support significant plant life in many areas of the Earth.

Substantial glacial melting in East Antarctica between 28,000 and 18,000 years ago brought cold fresh water to the oceans' surface around Australia and the oceans rose about fifty feet. All glaciers rapidly melted between 18,000 and 12,000 years ago and oceans rose another 300 feet. Carbon dioxide was far more soluble in the colder water on the surface of the oceans. Consequently, there was a massive carbon dioxide transfer from the atmosphere to the oceans during and shortly after the glacial breakup phase. This left much less carbon dioxide in the atmosphere to support life.

The low atmospheric carbon dioxide effect is one of the keys that caused the massive changes to life during the last extinction. It restricted the areas on land where plants could survive, thus restricting the areas on land where animals and humans that ate the plants could live. This led to many changes in distribution patterns of vegetation and the large animal extinction that paleontologists have identified. There were probably many plants that became extinct that are far more difficult to identify.

Looking Ahead

To see how all this fits together we will take a detailed look at how the glacial cycle works, and how changes to the oceans, atmosphere, and land during glaciation relate to extinction. We will study the plants, animals, and humans in determining the cause of extinction. The proof is in the details.

Most of the extinction story is a look at what happened in the past 28,000 years. The effects of large increases in atmospheric carbon dioxide will be discussed. We will assess our current situation in relation to the earth, plants, and animals. Possible happenings between now and the next glacial period will be suggested. We will speculate how the next glacial period will affect life. Since we are now in the best of times as far as the glacial cycle is concerned, we can expect that life will be tougher as the glaciers come back.

Chapter 2. Transitions Relating to Extinction

Before the large animal extinction near the end of the last glacial cycle, the earth was a very different place than it is today. During glacial melting, the earth became very dynamic with transitions in land, oceans, atmosphere, and life.

Before glacial melting, sea level was about 350 feet lower and the oceans had about 3% less total volume than today. Only water and some gases had been removed from the oceans by the evaporation/precipitation cycle that diminished the seas and built up the glaciers. The salts or minerals dissolved in the oceans remained in place, so the mineral concentrations increased from 3.5% to 3.6%. The saltier oceans changed the buoyancy and in some cases the internal makeup of organisms. The glaciers of the last ice age had been in place for about 50,000 years and the average ocean temperature was colder than today. However, the surface of the oceans in the lower latitudes was warming up.

Exposed Land

With the ocean level 350 feet lower, the surface area of the land was greater with a reduction in the surface area of the oceans. The continental margins were exposed and some shallow seas today were then dry land. This made all continents larger than they are today. Numerous land bridges were uncovered which connected continents with islands and other continents.

The land bridge from Alaska to Siberia across the Bering Sea provided a migration route for plants and animals between the North American continent and the Asian continent in both directions. Plant migration is a very slow process with blowing seeds and spores moving the plant community. This land bridge also provided a possible migration route for humans from Asia to North America.

Another land bridge connected much of Indonesia to the Asian continent. The greater Australian continent included Tasmania and New Guinea in one large land mass. The Australian land mass extended to the northwest towards Indonesia so that a sea voyage of only sixty miles was needed to connect Australia and the Asian continent through Indonesia. Humans crossed this sixty miles of ocean from Indonesia to Australia about 60,000 years ago.

Parts of the South China Sea, East China Sea, and Yellow Sea were all dry land. Korea and Japan were connected with a land bridge. Taiwan and Mainland China were also connected. The Philippines had a much larger land area and most of the islands were part of one land mass.

In North America, Texas was bigger than ever as it extended out into the Gulf of Mexico. Florida was twice as wide as currently and the Bahama Banks were one large land mass. Off the North American coast much of the continental shelf was exposed. Ice covered much of it, to the north, possibly including the Grand Banks of Newfoundland. In South America the exposed eastern continental shelf extended almost to the Falkland Islands. The Caribbean Sea was far more restricted than at present.

In Europe, the British Isles were connected to the rest of Europe. The North Sea and Baltic Sea were covered with ice, along with Scandinavia. The Celtic Sea south of Wales and Ireland was above sea level. The Strait of Gibraltar was more restricted than it is today and the Mediterranean Sea was much smaller. The Black Sea had no outlet to the Mediterranean Sea.

GREATEST GLACIAL EXTENT

The greatest glacial extent is determined by evidence of the last glaciation in the field and by using indirect methods. This gives an estimate of the total area covered by ice during the last glacial cycle. This occurred at a very active time for glaciers when there was a great deal of melting in some areas, but glacial expansion in other areas. This indicates that the greatest glacial distribution probably did not occur at any one time. For instance, the peak expansion in North America probably did not occur at the same time as the peak expansion in Antarctica. The dynamics controlling glacial expansion and glacial melting are different for the polar glaciation of Antarctica and the lower latitude glaciation of North America. If the greatest glacial extent had occurred at one time, the ocean level would have dropped 400 to 450 feet instead of 350 feet. We will examine the peak distribution of glaciers.

In North America, a huge glacial dome was centered approximately on Hudson Bay in Canada, extending south into Illinois, Indiana, and Ohio and as far west as Alberta. Another glacial dome was centered in the Canadian Arctic Islands. There was a large ice sheet centered over the continental divide in Western Canada, extending across southern Alaska and onto the Alaskan Peninsula. These ice sheets were so large that they had about the same volume as the present day Antarctica ice sheet. South of these ice sheets, many areas of glaciation covered the mountainous regions of the western United States. Northern Alaska also contained glaciation not associated with any ice sheet. The glacial dome over Greenland was more extensive than today, while Iceland was totally glaciated.

In Europe much of the British Isles, the North Sea and Scandinavia, along with northern parts of Germany, Poland, and into Russia, were covered by one ice sheet. This merged to the east with an ice sheet over the Barents Sea in the Russian Arctic and the adjacent Russian land. There was another ice sheet over the Kara Sea and the land to the south in Asian Russia. There is some evidence of another Asian ice sheet on the continental shelf in the East Siberian Sea and the Chukchi Sea with little ice over the present day land of far eastern Siberia.

In the Southern Hemisphere the Antarctic ice sheet covered a larger area than today extending out over the present day continental shelf. Also the ice was much thicker in most places, increasing the ice volume. It has been estimated that there was 26% more ice in the Antarctic ice sheet than at present. Glaciers covered the southern half of Chile, southern Argentina, and the adjacent continental shelves. Glaciers covered about 30% of the South Island of New Zealand and 20% of Tasmania. There was minor mountain glaciation in Africa, Australia, and New Guinea.

In addition to glacial coverage, we should look at periglacial conditions, which occur in areas either covered by glaciers or areas of permafrost. With permafrost, the ground stays frozen all year round and in some cases is frozen thousands of feet deep. These areas are capable of sustaining glaciers if any are in place. Currently about half of Canada and 70% of Alaska have periglacial conditions. Today about 10% of the earth's land is glaciated and an additional 10% has permafrost. During the peak glacial distribution, about 27% of the land worldwide was covered by glaciers, plus some permafrost increasing the periglacial areas. About 10,000 years ago, after the continental glaciers had melted, only about 10% of the land surface was periglacial. This was mainly the glaciers on Antarctica and Greenland, with only minor areas of permafrost not covered with glaciers.

CHANGING PLANT DISTRIBUTION AND CHARACTER

The distribution and variety of plants prior to glacial breakup was much different than it is today. In North America south of the ice sheets the vegetation was more like northern Canada and Alaska today with boreal woodlands and tundra. Figure 1 shows the generalized plant distribution in North America before the extinction at the end of the last ice age. This is referred to as the Pleistocene vegetational pattern.

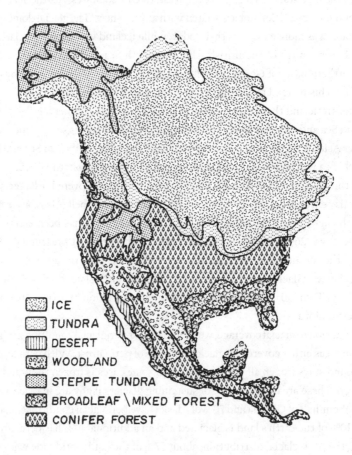

Figure 1. *Generalized vegetation formations of North America before glacial breakup (after Canby 1979 and MacDonald 1981 in Quaternary Extinctions).*

Much of Europe and Asia were covered by the steppe environment with patches of many mixed grasses and sedge plants interspersed with areas of tundra vegetation and shrubs. The sedges resemble grass but have solid stems instead of hollow stems. Tundra vegetation is usually found on permafrost ground and consists of lichens, mosses, and stunted shrubs. The steppe areas ranged from England through Europe south of the ice and through the northern two-thirds of Asia. These steppe areas also extended across the Alaskan-Siberian land bridge through Alaska and into the Yukon. This huge area is referred to as the mammoth steppe because mammoths were dominant throughout most of this area. There were also different mixes of plants in southern Asia than there are today, composed partly of plants now found further north.

Figure 2. Loess and sand dune deposits in North America formed near the time of glacial breakup — 18,000 to 10,000 years ago.

In greater Australia, including New Guinea, the boundary between tropical and temperate vegetation was about a thousand miles closer to the equator than it is today. Portions of southern New Guinea had temperate vegetation.

Much of the native vegetation in Australia today is unique, different from vegetation found anyplace else in the world. This is because Australia became isolated from all other continents at least 120,000,000 years ago. The animal extinction pattern in Australia suggests that most of these plants may have a greater atmospheric carbon dioxide requirement for survival than plants found in other areas.

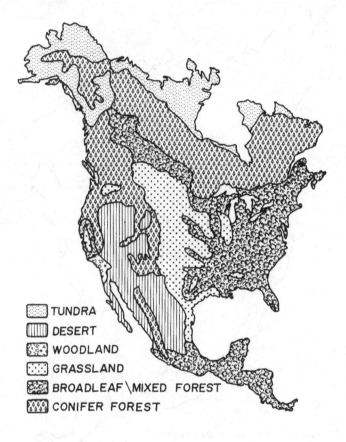

Figure 3. Generalized vegetation formations of North America today (after Espenshade 1970 in Quaternary Extinctions).

South American plants are also unique because South America was isolated from about 120,000,000 years ago until 3,500,000 years ago, when South America connected with North America. This isolated evolution over such a long time may have also created plants with a greater atmospheric carbon dioxide requirement.

Most animals have a fairly limited number of plants that they can normally eat and if these plants disappear, then the animals usually will also disappear. To compare, the vast human population eats parts of only about 3000 plants and of these probably less than 100 plants are significant staples.

The location of loess and sand dune deposits in North America are shown in Figure 2. Sometime between 18,000 and 10,000 years ago, most of the North American vegetation died out and loess and sand dune deposits formed on some of the bare ground. The global distribution of loess and sand dunes is found in chapter three.

The plant distribution in North America after the extinction is shown in Figure 3. This is the Holocene vegetational pattern, which developed when plants migrated in to re-populate the previously barren and ice-covered land. It is also today's vegetation distribution. There were new vegetation patterns in many parts of the world as plants migrated in to cover loess and sand dune deposits and bare ground. The climate and plant transitions led to animal transitions.

WOOLLY MAMMOTH AND RELATIVES

Before we talk about animals on each continent, it is worthwhile to examine the largest land animal. We will look at the worldwide distribution, prior to extinction, of one order of animals that includes the woolly mammoth and present day elephants. The order proboscidea had nine genera living before the last extinction and only one genus survives today, the modern elephant.

Many people marvel at these big critters with trunks and tusks, admiring their remains in museums, studying their life cycle, and also observing their only living relative, the elephant. One of my early memories was seeing the elephants in a circus parade. What really impressed me was that each elephant held the tail of the elephant in front of it with its trunk. While I was in college, studying geology in Wichita, Kansas, a woolly mammoth tooth was found in a nearby sand pit.

About twenty-five years ago the International Wildlife Park in Grand Prairie, Texas, received some twenty-five young orphan elephants from Africa. My wife and I went to see them and we were able to drive up to the fence of the holding pen and feed them. We had the passenger window rolled down and my wife was feeding an elephant one peanut at a time. It apparently wanted the peanuts faster and reached its trunk into the car window to search for them. In feeling around with its trunk in search of the peanuts, it also groped my wife. She threw the bag of peanuts at the elephant and I was instructed to drive away rapidly. I had the strange impression that this elephant had a sense of humor.

This also suggested to me that woolly mammoths may have been fun-loving critters. However, they were also one of the most successful mammals on earth before the large animal extinction. I always get a touch of melancholy when I think of the loss of the woolly mammoth and the other large Pleistocene animals.

The woolly mammoth had one of the widest ranges of any large animal in the late Pleistocene and is believed to have been the dominant genus throughout much of its range. Although it was dominant, it was not the most numerous, because of its large size. Early in the Pleistocene, about 1,700,000,000 years ago, the woolly mammoth migrated across the Alaskan-Siberian land bridge to North America. The woolly mammoth inhabited the grass lands in northwest North America covering parts of Alaska, Canada, and the lower United States. It was found as far south as Kansas and as far east as the east coast. In Europe the woolly mammoth lived on the mammoth steppe as far south as Spain and Italy. To the north it was in England and Scandinavia and to the east it lived in essentially all of Eastern Europe. The woolly mammoth's range in Asia included all of Russia, much of Kazakhstan, and parts of Mongolia and China.

One example of the abundance of woolly mammoths is an ancient human village of four dwellings made from the bones of many woolly mammoths that was located at Mezhirich, Ukraine. The biggest dwelling was made of almost twenty tons of mammoth bones and it included forty-six skulls, ninety-five jaw bones, forty tusks, and other bones. The bones are believed to have come from a nearby natural mammoth "cemetery," as the bones were not all dated as the same age. This village was occupied by people sometime between 15,000 and 14,000 years ago.

The woolly mammoth was about nine feet high at the shoulder, had a high domed head, with a topknot of thick hair, a humped sloping back, a short tail, and small ears. It also had large greatly curved tusks and a relatively short trunk ending in two fingers. It was covered with dense hair, a thick undercoat, and a layer of fat for insulation. Frozen and mummified woolly mammoth carcasses from Alaska and Siberia, along with cave paintings, have allowed us to know these animals pretty well. They were grazers, living on many species of grasses, sedges, and small shrubs. The Columbian mammoth, which was larger and had no hair, lived in the northwestern United States and as far south as Oklahoma.

American mastodons were found from Alaska to Florida, although they were most common in the eastern forests. They were browsers and fed on twigs, cones of conifers, leaves, coarse grasses, and swamp plants. They stood nine to ten feet at the shoulder and were more heavily built than mammoths. Mastodon tusks were higher than, and not as curved as, mammoth tusks.

The andine gomphothere, pampean gomphothere, and stegomastodon lived in South America. The andine gomphothere was especially common in the southern Andes. The pampean gomphothere was found mainly in Brazil and Ecuador. The stegomastodon was found in Brazil and Venezuela.

Before the last extinction, elephants had a larger range in Africa and Asia than they have had in the last 500 years. Now, let us look at the large animal life on the continents before the last extinction.

LARGE ANIMAL DISTRIBUTION

In North America there were forty-five genera of large (over one-hundred-pound) animals. Since about half of the continent was covered by glaciers, this means that the forty-five genera of large animals were occupying half the space where twelve genera now live. This great profusion of large animals were living at a time when the climate was colder than today. When the cold north wind blew in, it came over the glaciers, making it bitter.

Some North American animals were unique to North America; some were similar to animals found in Asia, Europe, South America, and Africa. Animals migrated over the Alaskan-Siberian land bridge and some came from South America through Central America to North America.

South America had a greater variety of large animals than any place on earth, with fifty-eight genera, but only twelve of those genera survived the extinction. Studies of continental drift indicate that South America split off from Africa and Gondwanaland and drifted away about 120,000,000 years ago. South American animals were an isolated population for over 115,000,000 years. During this time many strange and unfamiliar critters evolved and flourished. About 3,500,000 years ago South America and North America became connected by Central America. Many animals migrated through Central America in both directions giving a greater variety of animals to South America.

Europe, Asia, and Africa have been connected for many millions of years. Throughout this time there has been a great deal of migration of animals between the continents. Most of the time over the last 2,000,000 years there has been a land bridge between Asia and North America where animals migrated between continents. Consequently, many of the animal genera that lived in Europe and Asia also lived on other continents.

Of all the continents, Africa had the least number of animals that became extinct. Of forty-nine genera of large animals in Africa before the end of the last ice age, only seven genera did not survive.

The Australian continent is another matter altogether. Before the last extinction, there were twenty-two genera of large animals, but only three genera survived. Life in Australia in the Pleistocene was more difficult than on the other continents. The largest animals in Australia were only about one-third the size of the largest animals on the other continents. The twenty-two genera of large animals were much fewer than on any other continent. Some scholars believe that the poor soils and erratic climate restricted development of vegetation and of large animals. Those animals that developed had a smaller population density than large animals on continents with lusher vegetation. Extinction in Australia was not restricted to mammals, but included most of the large reptiles and many of the flightless birds.

Australian animals have been isolated since the breakup of Gondwanaland about 120,000,000 years ago. Evolution in Australia took a much different path than on all other continents. The marsupials became dominant in Australia and filled most ecological niches.

Throughout the world we had a great profusion of large and diverse animals before the last extinction. These animals were mainly mammals and were the product of 65,000,000 years of evolution since the dinosaur extinction. They were living off a plant distribution which was much different from today. Now we will examine the last extinction in greater detail.

Chapter 3. Extinction

This study indicates that low atmospheric carbon dioxide was the major cause of the large animal extinction near the end of the last ice age. There was not enough carbon dioxide in the atmosphere for most plants in the higher latitudes and higher altitude areas to survive. Some plants could not survive in the low latitude and low altitude areas. The reduction in carbon dioxide in the atmosphere occurred over thousands of years, and the dying off of the plants was a very gradual process.

Without sufficient plants to eat, most of the large animals could not survive. These large animals had been on earth for many millions of years and had survived many previous threats to their existence. Yet in a geologically short period of time they became extinct. We will now look at the sequence of events involved in the extinction.

Melting of the Continental Glaciers

In the coldest part of the glacial cycle, less heat came from the sun and the glaciers were fairly dormant. The ocean surfaces were cold so that not much evaporation took place, and there wasn't much moisture in the air, so not much snow fell onto the glaciers. The atmosphere was cold so very little melting took place.

As the heat energy from the sun increased, the atmosphere and top of the oceans warmed up. More glacial melting occurred and more water evaporated from the oceans, some of which fell as snow on the glaciers. The glaciers flowed faster and expanded where excess snow fell and dissipated where only melting occurred. Polar melting put cold water on the surface of the oceans close to the poles, which retarded evaporation. Warming oceans closer to the equator caused increased evaporation. This created a strange situation with significant melting of the polar glaciers of Antarctica between 70 degrees and the South Pole. Also, there was glacial buildup from excess snowfall on the glaciers between 40 and 70 degrees north latitude.

With the cold meltwater on the surface of the oceans in the high latitudes, the atmospheric carbon dioxide was being absorbed by the oceans, thus lowering the level of carbon dioxide in the atmosphere. The low atmospheric carbon dioxide affected Australia first about 25,000 years ago. There was melting on the huge East Antarctica ice sheet which put cold meltwater, undersaturated in carbon dioxide, around Australia. When the carbon dioxide in the atmosphere was reduced, it affected the Australian plants first. This in turn caused the large animal extinction in Australia long before it occurred on the other continents.

By 18,000 years ago, the glacial breakup was in full swing with substantial melting and little glacial recharge. Tremendous amounts of cold meltwater flowed into the oceans. Because the meltwater was fresh water, it was not as dense as the salty ocean water, so it floated on the surface of the oceans. Soon most of the ocean surfaces were covered with cold meltwater.

The cold meltwater on the surfaces of the oceans absorbed carbon dioxide from the atmosphere. Cold water can hold over twice as much carbon dioxide in solution as warm water. To put this into perspective, think about a can of pop. Coke, Pepsi, or 7-Up all have three basic ingredients: water, carbon dioxide, and sugar. The sugar does not have any effect on the solubility of carbon dioxide in water. If you open a cold can of Coke, nothing happens as the carbon dioxide stays in solution in the water. However, if you open a can of Coke that's been left out in the sun or in a hot car, it spews icky stuff all over. One very hot, thirsty day in West Texas I only had one hot Coke to drink. I carefully pointed it away from me and opened it slowly. My precautions did not help as it still blew almost all the Coke out and I ended up with just one small swallow left in the can. This is because a can of cold soda has about one can of carbon dioxide in solution and this comes out of solution fast when the soda is warmed and the pressure is released.

The cold meltwater in the oceans, which was under saturated in carbon dioxide, acted just the opposite as the super saturated hot Coke. It absorbed large amounts of carbon dioxide from the atmosphere.

Large Animal Extinction

As the glacial breakup continued, more carbon dioxide was removed from the atmosphere and this began to affect the plants at the high altitude and high latitudes. The plants at the margin could not grow to their full potential, and many stopped growing. An example of this occurs at the tree line today. In Alberta, Canada, just north of the Athabasca glacier, there is a stand of trees at the tree line that are over 300 years old but are only about six to eight feet tall.

Stunted growth of some plants is the first response to low atmospheric carbon dioxide. At the time of extinction, the trees near the tree line died and the tree line migrated down to a lower elevation and the Arctic tree line migrated toward lower latitudes. With less atmospheric carbon dioxide, all plants were less prolific, slower growing, and attained a smaller maximum size.

As the food supply dwindled, the animals had several responses. Initially the animal population thinned out and in areas where the food was very limited, adult animals were often smaller. Less land area was capable of supporting large animals and in some areas it disappeared altogether. When the food sources disappeared, the animals that ate that food either migrated or died.

In the eastern United States the boreal woodlands died out, which sealed the fate of most of the large browsing animals that lived in the forests. In western North America much of the grasslands and tundra died, which doomed most of the large grazing animals.

Many of the large North American animals that became extinct near the end of the last ice age are described below: the woolly mammoth, Columbian mammoth, mastodon, and cuviers gomphothere are all relatives of the modern elephant.

The stag-moose was about the size of a modern moose on steroids with long legs and complex antlers. It had a range from Alberta to the east coast. The large mountain deer ranged from Mexico to Wyoming. The stilt-legged deer lived in the east central United States.

The Western North American camel resembled the present-day dromedary, but had longer legs. It was a grazer and a herd animal. The long-legged North American llama was larger than the living South American llama and was a grazer. The stout-legged llamas lived in both northern South America and southern North America but did not survive.

The stock's pronghorn ranged from Mexico to Nebraska and the large four-horned pronghorn ranged from California to Mexico. The woodland muskox, a grazer, was found from Alaska to the east coast and as far south as Mississippi.

The shrub ox was a little larger than a muskox and ranged from California east to Illinois.

The horse was missing from the North American scene from its extinction about 10,000 years ago until the Asian horse was reintroduced by the Spanish in the 1500s.

The North American cheetah was related to the puma. It had long thin legs, a small head, and thin body. The American Pleistocene lion was much larger than the African lion and was found from Alaska to northern South America. The new world sabertooth was found throughout much of North and South America. It preyed on larger, slower-moving animals.

The megalonychid ground sloth was about the size of an ox, and lived in the eastern two-thirds of the United States and north to Alaska. The Shasta ground sloth was the smallest North American ground sloth, over 300 pounds, and lived from New Mexico to Alberta. The big-tongued sloth has been found at the La-Brea Tar Pits.

The giant beaver ranged from Alaska to Florida and was the largest rodent in North America. It was about the size of a black bear. The giant capybara lived in the southern United States and was substantially bigger than the current South American capybara.

The North American pampathere was similar to a very large armadillo and was three feet high and six feet long. It ranged from southeastern United States to Kansas. The North American glyptodont had heavy armor, was about four feet high, seven feet long, and weighed about a ton. It lived from Texas to South Carolina and browsed on tropical and subtropical vegetation.

Tapirs lived in North America until the end of the last ice age. Similar tapirs are still found in South America. The long-nosed peccary and flat-headed peccary are no longer with us. In all about 73% of the large animal genera in North America became extinct.

The survivors of the large animal extinction were those animals that could migrate to better areas and could adapt to changing conditions. The surviving bison and North American elk are examples of large animals that lived from the northern forests to the southern plains grasslands before the white man arrived in North America. All of the other large animals that survived the extinction, such as black and brown bears, mountain sheep, mountain goats, deer, moose, caribou, and pronghorn, had a very large and diverse range and were adaptable. This allowed them to find areas of favorable conditions where small bands of animals could survive.

The large animal extinction was more severe in South America as almost 80% of the large animal genera died out. This was a total of forty-six genera of very unique large animals passing into the great beyond. South America had barriers to migration such as the Andes mountains and Amazon River, which prevented many animals from migrating to better conditions. Some of the South American animals that became extinct are identified below.

The South American glyptodont was a large mammal with a heavy skeleton and massive body armor. It was similar to the North American glyptodont and lived in Venezuela and Brazil. The South American pampathere was related to armadillos and similar to the North American pampathere. It lived in Brazil. The South American bush dog had short legs and a short, broad face.

The macrauchenid had long legs, a long neck, and long snout with nose openings on the top of the skull. It was about the size of a camel and probably spent a lot of time in the water. The toxodont may have been the most common large South American animal near the end of the last ice age. It has been described as a giant guinea pig, built like a rhinoceros, with hippo-like habits. The two genera of South American horses also became extinct.

The hermit megathere was one of the largest ground sloths, with a length of twenty feet and a weight of three tons. It lived in South America, Central America, and North America. The megathere ground sloth was also about twenty feet long, but lived exclusively in South America. It was a browser and ate from trees while standing on its back legs.

The South American nothrothere was a smaller ground sloth. The big-tongued sloth was found in both South and North America. It had powerful arms and claws to dig out roots and was found from Chile to Columbia. The broad-faced sloth was large-sized and found from Argentina to Brazil. The mylodon was a medium-sized ground sloth found in southern South America. The scelidotherium ground sloth was found from Chile to eastern Brazil.

Many animals that became extinct in the higher latitudes of North America survived in the lower tropical latitudes of Central America and South America. These survivors include the tapir, spectacled bear, and some smaller cats such as jaguar, jaguarondi, and ocelot. In addition, some genera of sloth, capybara, peccary, and llama survived in South America after becoming extinct in North America.

In Europe and Asia, the woolly mammoth steppe environment did not survive and consequently neither did the woolly mammoth, dominant animal that it was, and neither did many of the large animals that had shared the environment. Some of the European animals that passed from the scene are enumerated as follows.

The European cave bear is known from the remains that have been relatively commonly found in caves. It was a very large bear with short legs and a big head.

The European cave lion was much larger than the living African lion. The striped brown and spotted hyenas lived in Europe and Asia until the end of the ice age and are still found in Africa. The Eurasian giant beaver was larger than the modern beaver.

The giant deer, also know as the Irish elk, was found in bogs from Ireland to Germany. The sabertooth lived in many areas of Europe and Asia. The European horse became extinct but the Asian horse survived. The giant rhinoceros was larger than any living rhinoceros and lived on the steppes of Eurasia.

The Asian genera that became extinct included the woolly rhinoceros, which was found in much of Asia and Europe. It was specialized to live in the cold, close to the glaciers. The steppe bison was a very common animal that ranged from England to Alaska. It had long horns, massive humped shoulders, and a slender rump.

The extinct Asian hog had some features found in older pigs. There was a water buffalo that lived in Palestine. The extinct Asian ox lived in South Asia and there was a nilgai which lived in East Asia. A form of waterbuck and an antelope from South Asia also became extinct.

The Eurasian extinction did not affect as high a percentage of large animal genera as the North American. This is because Asia extends almost to the equator and is bordered on the south by warm seas which gave off carbon dioxide. This made survival easier for plants and animals in the low latitudes. Many Eurasian animal genera migrated far into south Asia and survived. There were also a number of large animal genera that became extinct in Eurasia but survived in Africa. These African survivors included giraffe, lion, hyena, and leopard.

The African continent was least affected by the low atmospheric carbon dioxide because it straddles the equator and for the most part has low elevations. However, 14% of the African large animal genera did become extinct at the end of the last ice age.

Several of the animals that did become extinct are described as follows. Sivatheres were gigantic antlered relatives of giraffes. They had a massive body with normal neck and legs. They disappeared from Africa, Europe and Asia.

The alcelaphines, with long curved horns, were related to the hartebeest; they lived in east and south Africa. Another medium-sized alcelaphine lived in east Africa. The African long-horned buffalo was a gigantic horned buffalo which lived in north, east, and south Africa.

The Australian continent was a difficult place for large animals to survive in the Pleistocene. Before the extinction there were twenty-two genera of large animals in Australia. Only three genera survived and two of these were kangaroos.

We will look at some of the very unique animals that did not survive. The extinction was especially hard on large kangaroos, with seven genera not surviving. Those leaving the scene included the short-faced kangaroo and the giant short-faced kangaroo, which stood about ten feet tall. The largest extinct giant kangaroo was about eleven feet tall.

Four genera of diprodonts did not make it. One of these was the largest marsupial known at over six feet tall and ten feet long. Two genus of wombats became extinct, one of which was a giant wombat that was about as large as a boar hog. The marsupial lion was believed to be a carnivore and was about the size of a leopard. Other large carnivores were reptiles including a twenty-foot-long snake and a ten-foot-long land crocodile. Such large reptiles suggest a very warm climate over a long period of time.

The lack of diversity, high mortality of genera, and extinction starting at a earlier date indicate that the food source was far from adequate. Only very highly specialized animals could survive. It also appears that the Australian plants were very susceptible to lower atmospheric carbon dioxide and changes in climate, which many did not survive.

LOESS AND SAND DUNE DEPOSITS

When most of the vegetation disappeared because of low atmospheric carbon dioxide near the end of the last ice age, barren land remained. High winds over this barren land picked up silt-sized particles which created dust storms. When this windblown dust settled out of the air, it created large loess deposits. Some loess deposits are over 200 feet thick. The high winds also moved sand particles, creating large areas of sand dunes.

Most loess and windblown (eolian) sand dune deposits are loose fine-grained sediments which are rapidly eroded by water and wind. Consequently, most of these deposits are substantially smaller today than they were at the time of deposition from about 26,000 to 11,000 years ago. Also there is some evidence of smaller loess deposits that formed in the coldest part of the glacial cycle. This suggests that the atmospheric carbon dioxide was fairly low throughout much of the last glacial cycle, but it was not low enough to cause extinction until near the end.

The loess and sand dune deposits associated with the end of the last ice age are on every continent except Antarctica. The global distribution map (Figure 4) shows where loess and sand dune deposits are found and where loess and sand dune deposits are possible.

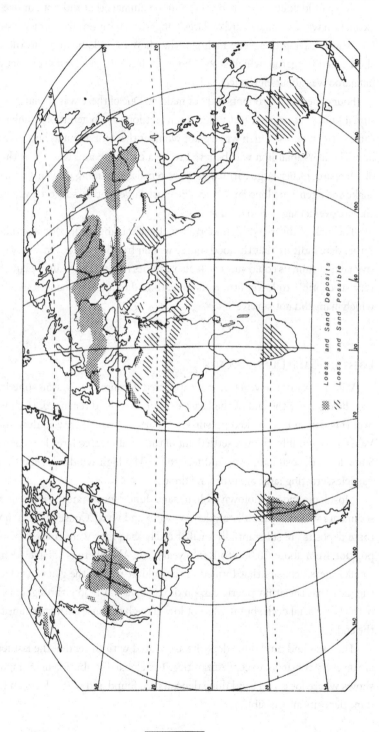

Major loess and sand dune deposits exist in the central United States from New Mexico to Ohio and from Louisiana to Minnesota. In South America, the loess and sand dune deposits are found in Uruguay, Paraguay, and Argentina. In Europe loess and sand dune deposits are found in west central and Eastern Europe, with the Ukraine in the center, extending into surrounding countries, especially Russia. In Asia, major loess and sand dune deposits can be found in China. Most of these deposits indicate areas where plants died from insufficient carbon dioxide in the atmosphere.

Dust and Sand Storm Effects on Animals and Humans

As the loess and sand dune deposits formed, the largest animals had to contend with more than just the loss of vegetation. The dust and sand storms had a debilitating effect on the animals. Most of the animals were weakened by insufficient food. When their lungs started to fill with dust, the end was near. The animals either migrated early in the dust storms or they died.

The large animals and plants that were decimated by the dust and sand storms had been food sources for humans. The dust storms were as destructive to humans as they were to other life forms.

In the North American plains, dust storms occurred in the 1930s that were not caused by low atmospheric carbon dioxide, although most of the dust storms were reworking ice age loess deposits. These dust storms were caused by drought, and deep plowing of the land. Deep plowing buries the vegetation and roots so deep that no binder is left on the surface of the soil to keep it from blowing away.

In the "1930s Dust Bowl," one of the major causes of death was "dust pneumonia." This was a lung disease caused by dust inhalation which affected the very young, old, and people with existing lung problems. I was a baby during the Dust Bowl and when the dirt blew, my mother covered my face with a wet handkerchief to keep the dust out of my lungs. She wrote a poem, "To a Kansas Dust Storm," at the height of the Dust Bowl in the. This may give some insight into what the people thought, about 13,000 years ago, before their demise in the ancient dust storms.

To a Kansas Dust Storm

> Out of the west, the dust witches come riding
> > riding the wind in dark chariots of dust
> with their black shirts, they put the sunlight in hiding
> > sweeping it clean, the very earth's crust.

> Man looks on in awe as the witches come swirling
>   sees desolate sand dunes where once there was sod
> their hate and defiance the dust witches hurling
>   is this grim destruction the vengeance of God?
>
> No, these are the monsters that man has created,
>   created from earth with the greed of his plow,
> so this is the vengeance of nature belated,
>   man reaps the reward of his heedlessness now.
>
> For gone are the glorious billowing grasses
>   that covered the prairie with carpets of green
> torn from the soil by the hurrying masses
>   seeking a fortune in grain's golden gain.
>   — Margaret Snook

The despair of those people about 13,000 years ago must have been overpowering as they choked from the dust storms. They had no idea why the dust storms were suffocating them. All they knew was that they were dying, along with their food sources.

The loess and sand dune deposits formed near the end of the last ice age are indicative of the last extinction which killed off many plants, animals, and humans. This extinction eliminated most of the large animal genera and deprived us of the chance to see this great variety of large animals.

At this point, we will examine how the earth works and the changes to the earth that led to the extinction. We will start by examining how the glacial cycle works.

Chapter 4. How the Glacial Cycle Works

The melting of the glaciers that led to the last extinction is part of a much bigger event, the glacial cycle. The glacial cycle causes changes in the composition of the atmosphere and oceans. It also causes major alterations to the surface of the land. These changing environments have a big impact on plants, animals, and people.

It is important to understand the forces that created the glacial cycles that have occurred throughout the last 1,900,000 years of the Pleistocene. When we see how the glacial cycle works, we can better understand the causes of the last extinction. We can also see why there were no extinctions at the end of the previous glacial cycles.

The glacial cycle is intimately associated with the heat input to the earth and what goes on in the oceans and atmosphere. Throughout the glacial cycle, glacial ice advances and retreats many times. These short term variations may take hundreds or thousands of years to complete. The last glacial cycle took about 100,000 years, and this is the model for the idealized glacial cycle presented later.

A common perception is that global warming and cooling is uniform. When the heat input is the lowest, the oceans and atmosphere are the coldest, with the glaciers at their greatest extent. Like many assumptions we read, this is not based on fact. The earth does not all cool down or warm up at the same time. It varies by hemisphere and latitude and is affected by a large number of factors. To get an idea of how heat input relates to temperature, let's look at the daily and yearly heat cycles in the North American interior.

Heat Cycles

During a clear day, the heat of the sun is most intense when it is directly overhead, so the greatest heat input occurs at noon. This is why they tell you not to try to get a suntan between 10:00 AM and 2:00 PM — because it toasts you, not tans you. However, the highest temperature of the day usually occurs between 3:00 and 5:00 PM, because the heat accumulates. Air and ground temperatures continue to build after the highest heat input. Only when the heat input is drastically reduced, as the sun gets lower in the sky, does it start to cool down.

The yearly cycle also has a cumulative heat lag with the seasons. In the northern hemisphere, the longest day of the year and the greatest heat input from the sun comes about June 21. However, the hottest weather of the year usually comes between July 15 and August 15. Also, the shortest day of the year and the lowest heat input from the sun occurs about December 21. The heat deficiency is also cumulative so the coldest weather occurs about a month later. That is why old people "snowbird" to the south in January.

During the yearly cycle the sun heats up the atmosphere and the land surface. However, most of the heat energy goes to heating up the top part of the oceans, surface water, ground water, and changing the phase of water from ice to liquid and from liquid to vapor. Since water takes a lot of energy to heat up and is slow to cool down, there is a lag from maximum heat input to maximum ocean surface temperature.

The heat buildup in the glacial cycle is similar to the heat buildup in the daily and yearly cycle. To make it simple we will apply seasonal terminology to the glacial cycle. The greatest heat input, or start of the glacial summer, occurs at about the start of the interglacial period. The high heat from the sun not only heats the atmosphere, the land, and the surface of the oceans, but causes a top-to-bottom heating of the oceans. It is this top-to-bottom heating of the oceans that causes the massive heat buildup between the point of highest heat input and the warmest point in the glacial cycle for the oceans. The warmest point for the atmosphere and the land occur substantially before the warmest point for the oceans.

At the start of the glacial winter, the heat input to the earth is the lowest in the cycle. The top-to-bottom cooling of the oceans causes a lag so that the oceans are at their coldest point substantially later than the lowest heat input.

Most continental glaciers break up when the heat input to earth is near its highest and the atmosphere is very warm. During the glacial breakup, the surface of the oceans become very cold because of the rapid influx of large volumes of fresh glacial meltwater and ice, both of which float on the saltwater of the oceans.

Energy for the Glacial Cycle

The glacial cycle is driven by energy. Vast amounts of energy are required in the oceans for glacial ice to accumulate. This energy evaporates water from the oceans, which is then carried in the atmosphere to the glacial site and deposited as snow and ice. The oceans receive 85% of the rain and snow that falls. Of the 15% that falls on land, much moisture falls over land areas that are not cold enough to sustain glaciation. Therefore, the oceans' energy requirement is many times that needed to evaporate the top 350 feet of the oceans. This accumulation phase takes warm oceans, a cool atmosphere, and about one-third of the glacial cycle. A present day example of this process is the lake effect snows around the Great Lakes in the late fall and early winter. Here there is heat energy in the Great Lakes left over from the summer, evaporating water. The cold atmosphere takes this moisture from the warm lakes and drops it on the surrounding area.

For glacial breakup to occur, vast amounts of heat energy are required from the sun and in the atmosphere to melt the glaciers.

The heat energy that fuels the glacial cycle comes from three sources: external energy from outside the earth, internal energy from the earth's interior, and surface energy generated on the surface of the earth. Of course, these three sources also provide all energy for the earth, but it is the variable energy input that causes the glacial cycle.

Almost all of the external energy reaching the earth comes directly from the sun. The amount of energy reaching the earth from the sun is variable for two reasons. First, the sun's energy output is not constant. Second, there are astronomical differences such as variable distance from the sun to the earth and the changing tilt of the earth's axis, which alter the amount of energy reaching the earth and how the energy is utilized.

Short-term variations in the sun's energy output have been measured. Sunspot cycles of about twenty-two years give changes in the energy output. Sunspots are large electro-magnetic storms on the surface of the sun which, with the associated solar flares, expel vast amounts of energy, some of which reaches the earth. The Maunder minimum in sunspot activity, when no sunspots were observed between AD 1645 and 1715, is believed to be part of the cause for the Little Ice Age which lasted from about AD 1450 to 1850. Polar reversals of the earth's magnetic field suggest very long-term variations in the energy output of the sun, which have not been identified. The last one occurred about 700,000 years ago.

Three astronomical variations affect the amount of energy reaching the earth and how this energy is utilized. First the earth's orbit around the sun is not uniform. The earth is 94,452,780 miles from the sun at its maximum distance and 91,342,080 miles from the sun at its minimum distance. The surface area of two spheres was calculated. One had the radius of the maximum distance of the sun and the other had the radius of the minimum. There was a 6.5% difference in surface area between the two spheres. This means that a square mile of earth receives about 6.5% less energy when it is at the farthest distance from the sun, compared to the shortest. This variation in the earth's orbit repeats every 93,000 years.

Second, the earth's axis of rotation is tilted at 23.5 degrees from vertical in relation to the orbital plane around the sun. This tilt changes; it ranges from 22 to 25 degrees. It takes about 41,000 years to make a full cycle. When the tilt is 25 degrees, the poles get more sun and are warmer. Less heat builds up at the equator because there is less sunlight coming in perpendicular. When the tilt is 22 degrees, the poles are colder and it is warmer at the equator.

Third, the gravitational effects of the sun, moon, and other planets cause the earth to wobble on its axis like a top. This wobble is unrelated to the changes of the tilt on the axis of rotation. One wobble takes about 21,000 years and the procession of equinoxes is one effect of this wobble.

The variable energy from the sun and the astronomical variations that affect it are by far the most important energy sources for earth and the glacial cycle. However, there is some effect from the earth's internal sources.

The earth has several internal heat sources. Radioactivity from elements such as radium, radon, thorium, and uranium generate considerable heat. Heat from impact of meteorites, comets, and asteroids were very important in the initial heat buildup. We also get heat from friction as the plates of the crust move against each other and from deeper movements in the earth. There is heat of compression from gravity forces. There are chemical reactions such as oxidation that give off heat deep in the earth. There are probably more heat sources that have not been identified.

A simple experiment suggests that there could also be considerable electromagnetic heating of the interior of the earth. In a seventh grade shop course we took an iron bar and wound copper wire around the bar. Then we put our bar on a stand and hooked the copper wire to a six-volt battery so we had an electromagnet. It attracted nails, iron filings, and other iron stuff. However, at the end of the day when I disconnected the battery and was putting up the electromagnet, I made a painful discovery. An inefficient electromagnet generates a lot of heat.

The iron-nickel core of the earth in some ways acts like a large bar magnet, thus creating the earth's magnetic field. However, in other ways the earth's magnetic field acts like a large inefficient alternating current electromagnet powered by the electromagnetic forces of the solar wind. Significant power is generated in the atmosphere by the interaction of the solar wind and the geomagnetic field. The aurora borealis is a visible energy component of this interaction.

During the aurora borealis, electro-potentials deep in the earth are altered. These spontaneous potentials, for correlation of formations, are measured in wells being drilled for oil. However, an aurora borealis totally disrupts these measurements in the northern United States and Canada. The solar winds interact with the magnetic field on earth and the disrupted spontaneous potentials suggest that the effect of the solar winds goes deep into the earth. From this it is suggested that the solar winds may provide power for the magnetic field. The strongest magnetic field in the last several thousand years occurred during the Maunder Minimum of sunspot activity. The strength of the magnetic field has diminished as the geomagnetic storms increased on the surface of the sun.

The magnetic poles wander from six to nine miles per year. Also there are magnetic pole reversals from time to time. In these cases the North and South Magnetic Poles change places. This is much like changing the polarity of an electromagnet. The last pole reversal occurred about 700,000 years ago.

This was discovered when pole reversals were found in paleomagnetism in rocks that had been molten when deposited. As the rocks cooled and solidified, magnetic minerals in the molten rocks aligned themselves with the magnetic poles at the time. After earth's poles reversed these magnetic minerals were pointing to the wrong pole.

Where the earth's magnetic poles are located today suggests that the iron-nickel core of the earth is not acting as a large bar magnet. On a longitude basis, they are about 120 degrees apart instead of 180 degrees apart to be on opposite sides of the earth. On a latitude basis, they are 142 degrees apart. Let us hope that the strange location of the magnetic poles is not a precursor of a magnetic pole reversal.

If the electromagnetic force of the solar wind is the energy source of the earth's magnetism, then changes in the sun's energy output probably causes the magnetic pole reversals. In addition, if the earth acts like a large inefficient electromagnet, this could indicate a much greater internal heat source for earth.

Internal heat gets to the surface of the earth by conduction, volcanic activity, and hot springs. Every place on earth has a thermal gradient with depth, indicating a heat release by conduction. This thermal gradient is variable from place to place. In much of the central United States the gradient is about one degree Fahrenheit per 100 feet of depth. However, in areas there magma bodies are close to the surface, like Yellowstone National Park, we get thermal gradients as high as one degree Fahrenheit per foot of depth. The gradual heat release to the surface by conduction is variable due to the rock type, rock fluid content, and proximity of the heat source to the surface.

Volcanic activity and hot springs release considerable amounts of heat to the surface of the earth. Linear volcanoes of the mid-ocean spreading zones, located at the plate boundaries of the earth's crust, create a disproportionately large amount of volcanic activity under the oceans. The best known of these mid-ocean spreading zones is the Mid Atlantic Ridge on which Iceland is located. There are about forty hot spots on earth, such as Hawaii and Yellowstone that give off lots of heat through conduction and volcanic activity. Variations in the frequency of volcanic activity on land areas of earth suggest that the internal heat sources may be variable.

In addition to external and internal energy sources, there is also surface energy generated by humans and natural causes. Lightning, natural forest fires, and chemical reactions, such as oxidation, that give off heat are examples of natural causes. Human activity that gives off heat includes electrical power, heating, explosives, internal combustion, nuclear heat, and even body heat. While the amount of energy generated on the surface of the earth may seem very large to humans, it is statistically insignificant compared to external and internal sources.

How Heat is Utilized on Earth

Heat energy from all sources heats up the oceans, atmosphere, and land. However, heating and changing the phase of water uses most of the heat energy. It takes more heat to increase the temperature of water than it does for any other naturally occurring substance except ammonia. One calorie of heat will raise the temperature of one gram of water one degree centigrade. This is called a specific heat of 1 and is the standard against which all other substances are measured. Most rocks require only about one-fifth as much heat as water to raise their temperature. The specific heat of granite is 0.192, quartz is 0.188, and marble is 0.21. Ice has a specific heat of 0.53. In the atmosphere, water vapor has a specific heat of 0.48, oxygen 0.218, and nitrogen 0.248. Not only does it take a lot of heat to raise the temperature of water, but there is a lot of water to be heated.

Water is an important, but small, component of the atmosphere in the form of water vapor, water droplets, and ice crystals in clouds. The average water vapor content of air is about 1.4% and it can go as high as 4%. Water is also the major component on the surface of the earth. Oceans cover 70% of the earth's surface, and they contain about 97% of the water in the water cycle. About 3% of the water on earth is fresh water. Close to 2.5% of the total water is in continental and mountain glaciers. Surface water on land, such as rivers, lakes, ponds, runoff, and fresh ground water, accounts for 0.75 % of the total water.

Water is the only naturally occurring substance that is found in all three phases: solid, liquid, and gas. It takes large amounts of heat to change the phase of water. It requires 80 calories of heat to melt one gram of ice. It requires 575 calories of heat to evaporate one gram of water.

About 30% of the energy coming from the sun is reflected back into space. The rest of the energy is absorbed by the atmosphere, oceans, and land. The reflectivity of different surfaces is called albedo. The albedo of various surfaces is shown in Table 1.

Table 1. Albedo of Various Surfaces.

Surface	Albedo
Clouds	30-90%
Snow	50-95%
Bare Ground	10-40%
Grass & Crops	20%
Forests	10-15%
Concrete	20-25%
Asphalt	8-10%
Water	7-10%

Albedo is the amount of the sun's energy that is reflected. The amount of energy absorbed is 100% minus the albedo. With clouds, some energy passes through and is neither reflected nor absorbed.

We can see from the albedo table that water reflects only 8% of the sunlight and absorbs 92%. With oceans covering 70% of the earth's surface, they store a tremendous amount of energy. This energy heats up the ocean and evaporates water from the surface.

On the land surface, up to 35% of the soil and rock volume is composed of void space. This void space between the grains or in cracks is often filled with groundwater. Much of the heat energy that is absorbed by the land is used to evaporate water from these void spaces, which moderates the land temperature. In the desert where there is no water in the voids, the surface of the land and overlying air heats up and cools down rapidly.

We can see that most of the heating by direct sunlight is mainly a water heating relationship. This energy heats up and changes the phase of water in the atmosphere, on land, and in the oceans. Large energy transfers are very important to the formation and destruction of glaciers.

How a Glacier Works

A glacier is a large mass of ice and rock that stays frozen year around. As it gets larger, it moves by the force of gravity, usually down slope. The glacier is formed by freezing water in place and by the compaction, melting and refreezing, and recrystallization of snow into ice. As the ice mass moves, it picks up rock from the earth's surface, which becomes part of the glacier.

The formation, location, and size of glaciers are dependent upon four factors, which are listed in order of importance. The first is land distribution on earth. This is where the land for the glacial site is located in relation to the North and South Poles and in relation to the oceans. The North and South Poles receive the least amount of energy from the sun and are the coldest areas. The land distribution in relation to the oceans is for water supply to feed the glaciers.

The second factor is water supply. Glaciers require vast amounts of water in the form of snow and ice. Usually this water comes from a nearby ocean source. It is delivered to the glacial site as water vapor and clouds in the atmosphere and falls on the glacier as snow and ice.

The third factor is temperature. The temperature has to be low enough in the glacier and in the atmosphere over the glacier so that melting and wastage does not exceed new snow and ice formation. Wastage refers to dissipation of the glaciers and icebergs breaking off and floating away. The oceans have to be warm enough for evaporation of large amounts of water.

The fourth factor is elevation. As moisture-laden air moves up slope over the glacier, it cools and snow falls. In addition, in higher elevations the thin air has lower heat transfer properties which retard melting.

Types of Glaciers

There are three kinds of glaciers: continental, mountain, and tidewater. Continental glaciers are formed on large land masses and the adjoining continental shelves. They are so large that once they are established, they generate most of their own climate. It takes a major climatic shift to breakup continental glaciers.

Although the earth is currently in an interglacial period, there are still two continental glaciers existing today on Antarctica and Greenland. They became smaller during the last glacial breakup phase, but remained intact. The most important factor in their formation and survival is land distribution. Both are located in polar or high latitudes and are surrounded by oceans.

Mountain glaciers may be found almost anywhere in the world in the higher elevations that have sufficient water supply. There are mountain glaciers near the equator on Mount Kilimanjaro in Africa, in South America, and other places. Although some mountain glaciers may be fairly constant or even growing, many have been receding for the last 200 years.

Tidewater glaciers terminate in the oceans. Tidewater glaciers are found at the margin of continental glaciers and some stand alone in high latitude areas. Tidewater glaciers in Alaska receded from about AD 1775 until about AD 1925, and today most are stable to advancing. It is believed that most of the mountain and stand alone tidewater glaciers today are remnants left over from the Little Ice Age.

It appears there were two main causes for the Little Ice Age. First, there was a greater warming of the oceans prior to the Little Ice Age. This put more energy into the oceans which enhanced evaporation and put more moisture into the atmosphere. Second, the Maunder Minimum in sunspot activity, which occurred from AD 1645 to 1715, reduced the sun's energy output. There was probably reduced sunspot activity prior to AD 1645 and after AD 1715 which caused the Little Ice Age to last from about AD 1450 to 1850. The start of the Little Ice Age is in dispute. Some people think it started earlier. These lower temperatures on earth condensed excess moisture out of the atmosphere in the form of snow and rain. During this time pre-existing mountain and tidewater glaciers advanced and more were formed. Now that the heat input to earth has increased after the Maunder Minimum, many mountain glaciers are still receding.

Currently, glaciers cover about 10% of the land surface on earth. During the peak of the last great ice advance, about 27% of the earth's land surface was covered by ice. Familiar places like New York, Chicago, Edinburgh, and Berlin were all covered by glaciers.

While glaciers cover only about 10% of the land surface today, another 10% of the land has permafrost, giving 20% as periglacial. Periglacial conditions are found over about 70% of Alaska and 50% of Canada. Most of the periglacial areas are in the Arctic desert environment and receive less than ten inches of precipitation per year.

Glacial formation requires large amounts of precipitation. However a small portion of the glacier may come from freezing either fresh water or salt water in place. Glaciers may build to a depth exceeding 12,000 feet. The ocean needs sufficient energy for significant evaporation in order to put large amounts of water vapor into the air moving over the glacial site. The air has to be warm enough to carry large amounts of water vapor. As the air rises over the glacier, it cools and drops its water as snow and ice onto the glacier. This process needs warm oceans with a cool, but not super cold atmosphere.

The water in the air is carried in two ways. First, very small water droplets and ice crystals are carried in air as clouds and fog. When the water droplets and ice crystals grow and become too heavy to be carried in the air, they fall as precipitation. In addition, water is carried as water vapor in air. When the air cools, the water vapor condenses into water droplets or ice crystals in the clouds or it can fall directly as precipitation.

Air's carrying capacity of water diminishes as the air density decreases. This can be with an increase in altitude or as you get closer to the North and South Poles. The water carrying capacity of air drops rapidly as the air temperature cools. Air at forty degrees Centigrade can carry 12.5 times as much water vapor as air at zero degrees C. Air at minus fifty degrees carries almost no water vapor. These are demonstrated in

Table 2. Absolute Humidity.

Temperature	Absolute Humidity
	Grams/Kilogram
40 degrees C.	47.3
30 degrees C.	26.9
20 degrees C.	14.7
10 degrees C.	7.67
0 degrees C.	3.80
-10 degrees C.	1.79
-20 degrees C.	.78
-30 degrees C.	.31
-40 degrees C.	.12
-50 degrees C.	.05

Table 2. Absolute Humidity refers to the amount of water vapor that can be carried by air at different temperatures at sea level. This is expressed in grams of water per 1,000 grams of air.

There is no direct evidence for much that has occurred in the glacial cycle. As has been indicated, there are many variables in the glacial cycle and they never repeat in exactly the same order. Despite these variables, it is worthwhile to construct an idealized glacial cycle. It will help us to understand better how the glacial cycle works. The idealized glacial cycle is based mainly on the last glacial cycle. For simplicity, a cycle length of exactly 100,000 years and seasonal terminology is used. The point of greatest heat input is the start of the glacial summer and the least heat input is the start of the glacial winter.

Idealized 100,000 Year Glacial Cycle

In creating an idealized glacial cycle, it must be understood that this is theory. There are many projections of the glacial cycle, on when the ice advanced and retreated, and the relative world temperatures. Some of these are based on: field mapping and dating of the distribution of the ancient ice sheets, climatic fluctuations interpreted from Greenland and Antarctica ice cores, oxygen-isotope analysis of pelagic foraminifer in ocean cores, beetle faunas in Britain, evolution of vegetation in Europe, and others. They all show the end of the last ice age at around 12,000 years ago, and that is where the good agreement stops. It gets fuzzier the farther back in time you go.

The glacial cycle is composed of a glacial period and an interglacial period, with the glacial period broken down into four phases. The interglacial period is the time of least glacial activity. At present, we have been in the interglacial period about 12,000 years and we may have about 6,000 years to go. The accumulation phase of the glacial period is the time when most of the glacial buildup occurs and lasts about 32,000 years. The dormant phase occurs when the higher latitudes on earth are the coldest and have little net buildup or dissipation of the existing glaciers and it lasts about 25,000 years. The active phase is a time when the glaciers are warming and there is more glacial movement. There is much more precipitation, and some of the glaciers are getting larger. At the same time, there is increased melting and wastage and some of the glaciers are getting smaller. During this phase, the glaciers reach their maximum extent and it lasts about 19,000 years. The glacial breakup phase is the shortest at about 6,000 years, but it creates the most change.

The accumulation and dormant phases are based more on theory with less available evidence. The active phase, breakup phase, and interglacial period have much more field data to substantiate them.

We will examine each part of the glacial cycle separately. Figure 5 gives a graphic representation of the idealized glacial cycle. We begin by looking at the interglacial period at about the start of the glacial summer.

Figure 5. Idealized 100,000 Year Glacial Cycle.

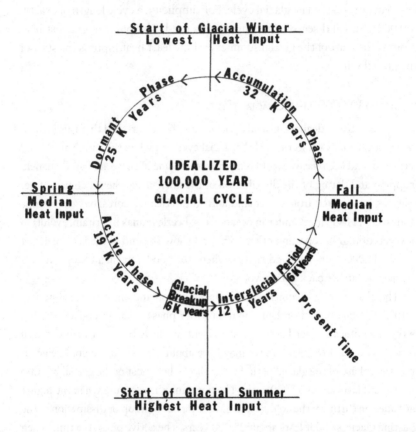

Time proceeds in a counter-clockwise direction, showing the interglacial period and the four phases of the glacial period. Heat input is indicated by the seasons.

INTERGLACIAL PERIOD

The interglacial period begins immediately after glacial breakup has occurred. At this point, the heat output to earth is the highest in the cycle. At the beginning of the interglacial period, the surface of the oceans is very cold. In the previous 6,000 years, most of the glaciers have melted and an influx of ice and very cold meltwater has raised the level of the oceans about 300 feet and lowered the surface temperature of the oceans dramatically. Initially, the atmosphere is relatively cool. As the interglacial period continues, the highest heat input of the cycle warms the atmosphere. At first, the precipitation is very low because the cold oceans do not put much moisture into the atmosphere. In addition, the carbon dioxide content of the atmosphere is near its lowest point.

The transition from glacial to interglacial takes some time to affect the oceans because of the vast amounts of heat required to warm the oceans. After the transition, the surface of the oceans heat up in a relatively short time. Once the surface of the oceans warm up, a more extensive top-to-bottom heating of the oceans begins, and the internal dynamics of the oceans change. The ocean currents become larger and faster.

More water evaporates from the oceans and precipitation increases. As the ocean surfaces warm, carbon dioxide comes out of solution and the atmospheric carbon dioxide level increases. The interglacial period is a time when life is good. Plants and animals flourish and move into land that was previously barren or covered with ice. The end of the interglacial period is not as clear cut as the start because there is a much longer transition into the accumulation phase.

Accumulation Phase

The accumulation phase starts in the late summer of the glacial cycle. Throughout the interglacial period there is a top-to-bottom warming of the oceans which continues, at a slower pace, in the early part of the accumulation phase. This warming stores a vast amount of heat energy in the oceans. During the accumulation phase, this energy evaporates the top 300 feet of the oceans and this moisture is deposited at the glacial site. The internal dynamics and current flow of the oceans are changed to bring warm water close to the glacial site. It also melts sea ice on the continental margins which allows warm water currents closer to the glacial sites.

The rise in atmospheric temperature slows during the early part of the accumulation phase in the late glacial summer. Near the start of the glacial fall, the atmospheric temperature cools. With very warm oceans putting large amounts of moisture into the air and a cooling atmosphere to condense the moisture out of the air, the weather gets colder and wetter. In the higher elevations and higher latitudes, this moisture falls as snow and the glaciers build. During the accumulation phase, new continental, mountain, and tidewater glaciers are formed, and all glaciers are advancing. When the oceans begin to cool, the atmosphere cools even faster and the cold weather, precipitation, and ice accumulation continue.

During the accumulation phase, the atmospheric carbon dioxide increases slowly until the surface of the oceans start to cool, and then it diminishes. The plants and animals that live where the ice starts to accumulate are in trouble. However, there is a transition time from the accumulation phase to the dormant phase so that the animals and most plant species migrate to areas with better conditions. In order to visualize the many facets of each phase of the glacial cycle, its worthwhile to develop a full data idealized 100,000 year glacial diagram.

Background Information for Full Data Diagram

There are many continuous changes that repeat with each glacial cycle. The most important of these changes are to ocean temperature, atmospheric temperature, carbon dioxide levels in the atmosphere, and precipitation. However, these changes are not uniform throughout the earth. These conditions are far more stable in the low latitude areas. The main reason is that areas near the equator receive a higher and more stable energy input from the sun. Consequently, the full data idealized 100,000 year glacial cycle diagram and most of the discussion refer to what is happening in the middle and high latitudes.

In the full data diagram we see that changes in each factor have a different timetable. For instance, the atmosphere reaches its highest temperature thousands of years before the oceans do. The timing for changes in atmospheric carbon dioxide is similar to the timing of changes in the surface temperature of the ocean, although not to the whole ocean. This is because the amount of carbon dioxide in the atmosphere is dependent on how much carbon dioxide is absorbed by the oceans. Carbon dioxide absorption is controlled by the surface temperature of the oceans. The heat energy in the oceans and the air temperature over the oceans and continents are all major factors in how much precipitation falls on the land.

On a global scale, weather patterns are quite variable. Many short-term changes in atmospheric temperature and precipitation are measured in years, decades, or hundreds of years. There are changes in ocean surface temperatures and atmospheric carbon dioxide levels that are measured in hundreds or thousands of years. Consequently, the changes during the glacial cycle are trend changes made up of many up and down changes. For instance, the slow increase in precipitation during the interglacial period we are in now is interspersed with times of drought. This gives a gradual increase in average precipitation over expanding areas, with occasional reversals. The temperature changes of the ocean refer to changes within the ocean, not just on the surface, except where specified. Next, we will look at the dormant phase.

Dormant Phase

The dormant phase begins at the start of the glacial winter, the point of lowest heat input to earth. The dormant phase has cooling oceans with slower cooling in the last half. The atmosphere initially cools rapidly, and then it maintains a very cold temperature. The carbon dioxide levels of the atmosphere drop as the ocean surfaces cool. This makes for low atmospheric carbon dioxide levels throughout the dormant phase. Much of the higher elevations and higher latitudes support little plant life because of the low atmospheric carbon dioxide. There is evidence of some loess deposits during this time confirming the reduced vegetation.

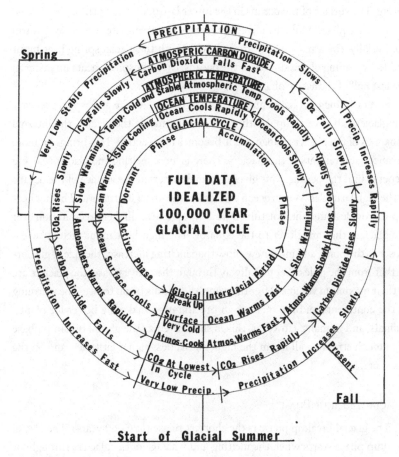

Figure 6. Full Data Diagram of Glacial Cycle.

When the oceans are cold, sea ice builds around the glacial areas cutting off new supplies of moisture. With both cold oceans and cold atmosphere, there is very little moisture in the air to form new precipitation. Without new snow and ice, the glaciers are more stable. As the temperatures inside the glaciers cool and more internal water freezes, the internal dynamics of the glacier slow down. These factors lead to very few advances and retreats of the glaciers, so they are fairly dormant. The glaciers become more mobile in the active phase.

Inner circle shows glacial cycle. Second circle shows ocean temperature. Third shows atmospheric temperature. Fourth shows atmospheric carbon dioxide. Outside circle shows precipitation. Time proceeds counter-clockwise.

Active Phase

The active phase starts with the median heat input to earth of the glacial spring. The surface of the oceans in the lower latitudes warms throughout most of the active phase while the lower portions of the oceans are very slow to warm up. Initially, the atmosphere warms slowly. With the glacial spring heat input and the warming of the surface of the oceans, the atmosphere heats up faster in the last half of the active phase.

Several things are happening with the glaciers. As the atmosphere warms, the glaciers also warm up. With more melting inside the glacier, the internal dynamics of the glacier change and it becomes more active. This brings on more advances and retreats of the glaciers. There is more melting and glaciers calving icebergs into the oceans. This gives cold meltwater on the surface of the oceans in the high latitudes. Warm ocean surfaces in the lower latitudes and higher heat input provide abundant moisture in the atmosphere for higher precipitation over the glaciers that are closer to the equator. The high heat input melts glaciers closer to the poles, and more new snow than melting increases the size of glaciers farther from the poles. About halfway through the active phase, the glaciers are at their maximum volume. After this time, the oceans begin to rise. The warming of the continents and the increase in precipitation make life better for plants, animals, and humans. However, this is short-lived as the glacial breakup phase puts much stress on all life in the higher altitudes, higher latitudes, and on the sea shore.

Glacial Breakup Phase

The glacial breakup phase is the shortest phase (at 6,000 years). The glacial breakup phase starts when the melting and wastage of the glaciers throughout the world greatly exceed new snow and ice, and the glaciers start to recede rapidly. At the time of the glacial breakup, the surfaces of the oceans are warm in the low latitudes but cool in the higher latitudes. The atmosphere is warm, and the heat input to earth approaches the highest point in the cycle. Also, initially the atmospheric carbon dioxide levels fall because of the cold meltwater on the ocean surfaces in the high latitudes.

As the melting continues, the oceans rise at a much faster rate, and before the glacial breakup is complete, the oceans rise about 350 feet. The glacial meltwater is very cold, which rapidly cools the ocean's surface. With the rise in sea level, glaciers and ice sheets on the continental margins are floated off the sea floor and drift out into the oceans. As the glaciers become warmer, they flow to the sea faster and calve icebergs. Much of this ice, along with cold meltwater, flows to most parts of all oceans, making their surfaces very cold.

The cold ocean surface does not have sufficient energy to evaporate much water into vapor, so the atmosphere is relatively dry. The atmosphere is warmer than the oceans, from the high heat input. Melting the glaciers uses most of this heat. There are indications that large amounts of dust were deposited on many glaciers in the last breakup phase. This dust increased the energy absorption of the glaciers, producing faster melting which accelerated the breakup. With a relatively warm dry atmosphere over the continents, very little precipitation fell to re-supply the glaciers.

Carbon dioxide is about twice as soluble in cold water as it is in warm water, so the cold ocean surfaces take carbon dioxide out of the atmosphere and into solution in the oceans.

The idealized 100,000 year glacial cycle represents what happened during the last glacial cycle. The next glacial cycle should have all of the five components described, but the length and magnitude of each phase may be quite different. There are many variables involved in the glacial cycle so that each cycle is different. However, all that we know now leads us to believe that there will be more glacial cycles. Next, we will take a closer look at what happens in the oceans leading to the large land animal extinction.

Chapter 5. Ocean Changes Relating to Glaciation and Extinction

Many ocean changes are caused by energy transfers. During an interglacial period, a very large amount of heat energy from the sun builds up and is stored in the oceans. Later this energy evaporates ocean water, some of which falls as snow and ice to form continental glaciers.

In the albedo chart (Table 1), we saw that only 8% of the sun's energy reaching the surface of the water is reflected. The other 92% of the sun's energy is absorbed. Of this energy, about half is absorbed in the top 33 feet of the ocean. The other half is absorbed from 33 feet down to the depth that light penetrates, close to 650 feet. The energy in the warmer waters heats the other parts of the oceans through turbulent flow and distribution by currents.

The top-to-bottom heating of the oceans is a slow process because warm salt water is less dense than cold salt water, and tends to float on top. Volcanic heat rising from the mid-ocean spreading zones increases circulation, but this is a minor source of heat compared to the sun. New continental glaciation begins when the oceans are near their warmest temperature, the solar heat input to earth is diminishing, and the atmosphere is cooling. As we have previously noted, it takes one calorie of heat to raise the temperature of one gram of water one degree centigrade. It takes 575 calories of heat to evaporate one gram of water. Therefore, to evaporate a gram of ocean water requires taking heat from the surrounding ocean. If there is little external heat added, it takes a lot of energy in the ocean to evaporate water from the surface and supply snow to the glaciers. New continental glaciers start to accumulate only when there is high heat energy in the oceans built up during the interglacial period.

Early in the last glacial accumulation phase there were warming oceans and a slowly warming atmosphere resulting in increased precipitation. Then a cooling atmosphere with relatively warm oceans extended the time of high precipitation. This provided the snow and ice for the glacial buildup.

Getting this moisture to the glaciers requires very dynamic oceans. The Gulf Stream and Japanese currents were much larger and faster-moving. Hurricanes and typhoons were bigger and carried their moisture much farther away from the equator when the oceans were warmer. Hurricanes and typhoons have always been major rain makers and when the glaciers formed, they were major snow makers when they reached farther into the higher latitudes of the world. Hurricanes have always required temperatures on the surface of the oceans in excess of 80 degrees Fahrenheit in order to form, regardless of latitude. Very large Nor'easters may have provided moisture to the eastern Canada glacial sites. Events like El Niño may have been much larger than today and affected a much greater area with a warmer ocean. All of these were pumping moisture to the glacial sites. With a top-to-bottom heating of the oceans throughout the interglacial period and well into the accumulation phase, we got these very warm oceans and their many effects.

As the sea level dropped, there was a contraction in the surface area of the oceans. Continental margins and land bridges were exposed, and some shallow seas dried up. The smaller area of the oceans changed the currents and circulation patterns. Circulation between the Pacific and Arctic Ocean stopped. Circulation between the Atlantic Ocean and the Mediterranean Sea was greatly reduced, and the Black Sea no longer had a viable ocean outlet.

Atmospheric Gases in the Oceans

The oceans have large amounts of atmospheric gases in solution. At times, some of this gas moves from the atmosphere to the oceans and from the oceans to the atmosphere. The solubilities of atmospheric gases in water are vastly different. In cold water, carbon dioxide is thirty-five times more soluble than oxygen, seventy-three times more soluble than nitrogen, and thirty times more soluble than argon.

The movement of oxygen, nitrogen, and argon has very little effect on the concentrations of these gases in either the oceans or the atmosphere. However, it is a different story with carbon dioxide. Since carbon dioxide is such a small component of the atmosphere, large changes in its concentration occur when significant amounts are absorbed by the oceans. The oceans contain about sixty times as much carbon dioxide as the atmosphere, so ocean concentrations change

Geologists know that carbon dioxide is absorbed in rainwater to form carbonic acid because this is a major weathering agent of limestone. Large areas called Karst Topography have many sinkholes and underground channels eroded out by water carrying carbonic acid. Caves in limestone, such as Carlsbad Caverns in New Mexico, were formed by carbonic acid dissolving out the limestone. There is also quite a bit of carbon dioxide exchange between the oceans and atmosphere by the action of waves at the surface of the oceans. Carbon dioxide is transferred from the oceans to the air when water is evaporated and carbon dioxide is released from the solution.

The carbon dioxide content of the oceans is affected by many processes other than the transfer between the atmosphere and oceans. Carbon dioxide is removed from the oceans and deposited on the floor of shallow seas and oceans within limestone and dolomite. This can occur from lime-secreting plants such as blue-green algae. It can also occur by the deposition of skeletal remains of marine animals. This can be external remains, such as shells, or internal bones and teeth of fishes and mammals. Some limestone and dolomite is chemically precipitated. The deposition of limestone and dolomite occurs at a slow rate in relatively shallow oceans. At depths below two to three miles with high pressure and cold water, no limestone deposition occurs because the carbonate stays in solution.

A large amount of carbon dioxide in the oceans is utilized by plants, taking in carbon dioxide and expelling oxygen. The carbon is often deposited as carbonaceous material in the sediments, or as a component of oil, natural gas, or coal deposits. Marine plants range from microscopic one-celled plants to forests of kelp that are as tall as many tree forests on land. The life processes on land and in the oceans utilize oxygen and carbon that is already in the system. Some of the oxygen and carbon is removed from the system and stored in rocks. However, life processes create no new carbon or oxygen.

Weathering may liberate large amounts of carbon dioxide from the rocks. However, the greatest source of new carbon dioxide in the oceans and the atmosphere is the original source, volcanism. This input of carbon dioxide into the system is spectacularly demonstrated in volcanoes above sea level, although the greatest amount of carbon dioxide is put into the system by underwater volcanoes. The mid-ocean rises are the site of sea floor spreading where two plates of the earth's crust are moving apart. The spreading gap is filled with new volcanic lava forming large linear volcanoes. In association with the lava deposition, there are releases of volcanic gases through fracture systems and at the lava deposition site. These linear volcanoes in the spreading zones have a combined length of very little. However, minor changes in solubility of the surface of the oceans create large changes of carbon dioxide concentration in the atmosphere.

All of the atmospheric gases are much more soluble in cold water than they are in hot water. For example, carbon dioxide is 2.4 times as soluble in water at 0 degrees Centigrade as it is in water at 25 degrees centigrade. This means that water on the surface of the Arctic and Antarctic Oceans can hold over twice as much carbon dioxide in solution as ocean water at the equator.

Cold water that is warmed in the tropics gives off carbon dioxide. The warm water currents on the surface of the oceans cool as they head toward the poles and they absorb carbon dioxide. There is a seasonal effect with carbon dioxide being given off in the late summer as the surface of the oceans warm up. In addition, carbon dioxide is absorbed in the late winter as the surface of the oceans cool down.

There is evidence from Alaska that shows the movement of carbon dioxide between the oceans and atmosphere. The evidence for carbon dioxide removal from the atmosphere by cold oceans is found in the tree lines. There are essentially no trees on the west coast of Alaska and the Aleutian Islands adjacent to the very cold waters of the Bering Sea. So much atmospheric carbon dioxide is absorbed in the cold Bering Sea that there is not enough remaining in the atmosphere over the adjacent land areas to allow trees to grow. There are abundant trees inland at the same latitude. Trees are found close to 2,000 feet in elevation in Denali National Park.

During the glacial cycle, the concentrations of carbon dioxide in the atmosphere change with the surface temperature of the oceans. Late in the interglacial period, when the surfaces of the oceans are warm, carbon dioxide comes out of solution, and the concentrations in the atmosphere rise. During the dormant phase, the surface of the oceans is cool. Carbon dioxide is absorbed by the surface of the oceans and the concentrations in the atmosphere drop. During the glacial breakup, the surfaces of the oceans are composed of cold meltwater which is undersaturated in carbon dioxide. Large amounts of carbon dioxide go into solution in the oceans and the atmospheric concentrations reach the lowest point in the cycle. The mechanisms for the transfer of carbon dioxide from the oceans and back are complex.

Most clouds are composed of a very large number of tiny water droplets. These droplets have an extremely large surface area through which carbon dioxide is taken into solution. As these droplets get together, they fall as rain. If the surface of the ocean is not saturated with carbon dioxide, then as the raindrops penetrate the ocean, the carbon dioxide in the raindrops stays in the ocean. Oceans receive 85% of the rain that falls on earth.

about 46,000 miles. Of course, these volcanoes are not all active at the same time. Sea floor spreading is a relatively slow process.

This volcanic activity is poorly understood, partly because we cannot see how it works. Most of the spreading zones are in the deep ocean and have been observed only by deep sea submarine research vessels. Basically, the only places where the mid-ocean spreading zones are above sea level are on Iceland and a few smaller islands. All of these volcanoes put out large amounts of carbon dioxide.

PRELUDE TO EXTINCTION IN THE ACTIVE PHASE

During the active phase, the surfaces of the oceans heated up rapidly, and the deeper portions of the oceans started a slow top-to-bottom warming. With this energy buildup there was increased evaporation.

At the same time, the temperature of the atmosphere was cool but rising rapidly. The warmer atmosphere warmed the glaciers causing internal melting. With the lubrication from the internal melting, the glaciers flowed faster. In the tropics, there was a rapid warm up of the oceans from the high heat input. Currents brought warmer water away from the equator and toward the poles. With more surface water evaporating from the oceans, and the warm atmosphere carrying this moisture to the glacial site, more glacial ice was formed. Therefore, at the time that new glacial ice was being formed, there was increasing flow of the glaciers with increased melting and wastage.

Halfway through the active phase, about 28,000 years ago, the melting and wastage was occurring on essentially all glaciated areas. This put cold meltwater on the ocean surface in the high latitudes. Closer to the equator, the ocean surfaces were still very warm and put lots of moisture in the air. Some of the moisture laden air deposited snow on the mid-latitude glaciers so that they grew and expanded.

The northern and southern hemispheres were evolving differently. In the northern hemisphere most of the glaciation occurred from 45 to 70 degrees north latitude. The southernmost glaciers in the northern hemisphere were expanding while the northernmost glaciers, such as those in Greenland and others bordering the Arctic Ocean, were contracting from melting and wastage. In the southern hemisphere, almost all of the glaciation was in Antarctica which fell between 70 degrees south latitude and the South Pole. Consequently, in the southern hemisphere there was melting and wastage of the glaciers and essentially no recharge.

The meltwater from Antarctica cooled the surface of the southern oceans so that there was little evaporation or precipitation. The cold meltwater on the surface of the southern oceans also removed carbon dioxide from the atmosphere. The lower atmospheric carbon dioxide caused the extinction to start in Australia during this time, which was earlier than on the other continents. Australia was surrounded by cold meltwater that came from the very large East Antarctica ice sheet.

The net effect to the oceans surface during the active phase was cooling in the high latitudes and warming near the equator. In addition, the ocean level rose from melting ice as much as fifty feet in the last half of the active phase.

Changes to the Oceans during Glacial Breakup

The final breakup of the continental glaciers at the end of the last ice age was rapid, taking only about 6,000 years. This process was extremely dynamic and greatly affected the oceans. The high heat input warmed the glaciers internally, increasing the flow and melting. Those glaciers that ended at the oceans' edge started calving very large icebergs into the oceans. The glaciers that ended on land melted and receded. Large amounts of very cold meltwater cooled the surface of the oceans.

The oceans rose at a faster rate as the glaciers melted and calved icebergs. As the oceans rose, very large glacial extensions on the continental shelves floated off the bottom. This brought super large floating ice sheets out into the oceans. This breakup of the glaciers on the continents was very rapid. The oceans rose about 300 feet in 6,000 years, which was an average ocean rise of one foot every twenty years.

The glacial meltwater, melted icebergs, and ice sheets were all fresh water. The fresh water was not as dense as the oceans' salt water and floated on the salt water, while mixing was relatively slow.

When I was in high school, one of my teachers told a true "follow-the-directions" story. In the 1600s two sailing vessels met off the coast of Brazil, out of site of land. The crew of one ship pleaded with the other for fresh water as they were dying of thirst. The other crew called back, "cast down your buckets." The first crew did not believe the directions and continued pleading for freshwater. Once again, the reply was "cast down your buckets." Finally, the thirsty crew did cast down their buckets and found that the water was fresh. They were opposite the mouth of the Amazon River and the fresh river water was floating on the dense ocean saltwater. I asked my dad, who had been a sailor on a destroyer during World War I, how the second ship knew they were opposite the Amazon River. He said when approaching port, the water changed color when the ship was still far out to sea, long before you could see the shore. This was very evident if you were standing watch in the crow's nest high above the water. The Amazon water carried sediment which made it a different color from the ocean water.

In a more recent example, during the 1993 floods of the Mississippi River, the huge volumes of freshwater coming from the mouth of the river floated on the surface of the Gulf of Mexico. This caused a ten mile wide river of freshwater which flowed on the surface through the Gulf of Mexico, around the tip of Florida, and out into the Atlantic Ocean. These two examples demonstrate that fresh meltwater floating on the surface of the rapidly rising ocean is a viable concept. In addition, it is a certainty that the surface of the ocean was much colder, even where some mixing had occurred. Where the meltwater flowed from the glaciers across the land and drained into the oceans greatly affected the oceans, atmosphere, and land.

Meltwater Distribution during Glacial Melting

During the melting of the last glacial cycle, about 20% of the ice volume that had caused lower sea level was in the southern hemisphere, most of it in Antarctica. This left about 80% of the ice volume causing lower sea level on the land areas of the northern hemisphere.

In the southern hemisphere early in the glacial melting, much of the meltwater and floating ice came from East Antarctica and went into the South Atlantic and Indian Oceans. This caused substantial early cooling of the oceans' surface around Australia. Later melting in Antarctica cooled the surface of the South Pacific Ocean.

In the northern hemisphere at the start of the breakup phase, meltwater and floating ice from the Cordilleran ice sheet in western North America flowed into the northern Pacific Ocean. This caused some cooling of the surface of the northern Pacific Ocean. Most of the northern hemisphere's meltwater and floating ice initially flowed into the Arctic and North Atlantic Oceans. This rapidly cooled the entire surface of the North Atlantic Ocean to near freezing temperatures. The level of all the oceans on earth was rising at an average rate of about one foot every twenty years. The Arctic and North Atlantic Oceans were rising faster than that. Because of this gradient from higher sea level in the North Atlantic to the lower level in the south, ice and meltwater moved south fairly rapidly. The very cold surface water of the North Atlantic Ocean very soon cooled the South Atlantic Ocean.

When the cold meltwater and some floating ice reached the lower South Atlantic, some may have moved through the Drake Passage between South America and the Antarctica Peninsula directly into the South Pacific. However, if today's current patterns indicate previous ocean currents, most of it went east, south of Africa and into the Indian Ocean. From there it went south of Australia and into the South Pacific. This means that all of the oceans of the earth were cooled to some extent early in the breakup phase by melting of continental glaciers throughout the world.

Probably less than halfway through the breakup phase, the land bridge between Asia and North America was breached. Cold water from the Arctic Ocean started coming through the Bering Strait. For most of the last half of the breakup phase, substantial amounts of cold meltwater and some floating ice flowed through the Bering Strait and into the North Pacific Ocean. This cooled the surface of the North Pacific Ocean a great deal near the end of the breakup phase.

Although most of the surfaces of the oceans were greatly cooled by meltwater and floating ice, there were some areas sheltered from the cold water that never cooled much. These warm seas brought pockets of rain and higher carbon dioxide levels in the atmosphere to the adjacent land areas. They included the Gulf of Mexico and the Caribbean Sea in the western hemisphere and the Mediterranean and Black Sea in Europe. The warmer seas adjacent to Asia were the Arabian Sea, the Bay of Bengal, the South China Sea, and possibly the Philippine Sea. Probably other warmer ocean areas have not been identified. However, the majority of the ocean areas still had very cold surfaces which had a major impact on the atmospheric carbon dioxide.

Carbon Dioxide Absorbed by the Oceans

The carbon dioxide level in the atmosphere was relatively low at the start of the breakup phase, but dropped rapidly. The cold, undersaturated surface water of most oceans absorbed vast amounts of carbon dioxide throughout the breakup phase of the glacial cycle. Near the end of the breakup phase, the carbon dioxide levels in the atmosphere reached their lowest point in the last 4,000,000,000 years at about half the present level or about 180 parts per million. This drop in atmospheric carbon dioxide meant that most plants could no longer survive in the higher latitudes and higher elevations, and some plants could not survive in the low latitudes. The animals that ate these plants either migrated to where the plants survived or they perished.

Because the surface of the oceans was very cold during the glacial breakup phase, there was not sufficient energy in the oceans for much evaporation. With reduced evaporation, there was also reduced precipitation over much of the earth. The tropics were not as affected as the rest of the earth, but precipitation dropped off rapidly going away from the equator. At the end of the breakup phase and the start of the interglacial period there was another dramatic change.

After the Glacial Breakup

The start of the interglacial period was also the start of the glacial summer. With melting of the glaciers essentially finished, the influx of meltwater to the oceans stopped. The energy of the highest heat input to earth was utilized heating up the oceans, atmosphere, and land. The surface temperature of the oceans began to warm rapidly and a top-to-bottom warming started. The temperature of the atmosphere rose rapidly with the warming oceans. As the temperature of the surface of the oceans increased, some carbon dioxide came out of solution, moving from the oceans back to the atmosphere. Worldwide precipitation increased slowly with the increase in temperature of the oceans and atmosphere.

The increased carbon dioxide in the atmosphere and increased precipitation during the interglacial period was great for life on land. The population of land plants and animals expanded, and they migrated away from the oceans' shores and to higher latitudes and higher elevations. Humans followed the plants and animals, and their population multiplied.

There is a very close relationship between the oceans and atmosphere, with the oceans being dominant. In this relationship, water and gases migrate freely both ways between oceans and atmosphere. There are energy exchanges between the oceans and atmosphere, but most of the exchanges are from the oceans to the atmosphere. Next, we take a closer look at changes in the atmosphere which relate to the large animal extinction.

Chapter 6. Changes in the Atmosphere during the Last Extinction

It is worthwhile to look at where the atmosphere came from and how it has evolved and changed throughout the earth's history. This helps to put the atmospheric changes during the last extinction into perspective.

It is believed the earth was formed by accretion. We can think of accretion as little chunks of space stuff sticking to a larger chunk, and as this got bigger gravity took over and pulled in more stuff. The earth increased in size with the addition of space dust, meteorites, asteroids, and comets. The early earth was probably hotter than today because of the much greater heat of impact and higher concentrations of radioactive elements which gave off heat.

Origin of the Atmosphere

Once the earth had sufficient mass for gravity to hold an atmosphere, then gases from the incoming space stuff added to the atmosphere. Some of the bigger pieces affecting earth melted, releasing some gases. This gas contribution continued as the earth rapidly reached close to its present size. Originally, much of the space stuff was traveling in essentially the same direction as the earth so the impacts would not be severe enough to release much gas. In addition, the atmosphere did not start to form until the earth was larger than the size of the moon today.

The main source of gas in the current atmosphere comes from de-gassing the planet. This has occurred throughout the history of the planet, as gravity segregation of the earth has brought light gases to the surface. Gravity segregation is the process by which heavy components of the earth are pulled toward the center and lighter materials float to the surface. This gives layers with a very heavy inner core at the center with a lighter outer core. There is a lighter mantel surrounding the core and a still lighter crust on the surface of the earth. On top of the crust are the oceans and glaciers, with a very light atmosphere above everything else. Most of the gas reaches the atmosphere through volcanic eruptions either directly into the atmosphere or into the oceans.

Carbon dioxide is found in large volumes in some hydrocarbon reservoirs. Most of it is found in natural gas reservoirs along with methane, but some is found in reservoirs that contain only carbon dioxide. The gas comes from deep inside the earth by way of natural fractures. It can also be associated with volcanism. For example, in New Mexico near the town of Clayton is a geologic feature called the Bravo Dome. At depth, a large reservoir in the Bravo Dome contains only carbon dioxide believed to have come from volcanism in the area.

Many very deep natural gas reservoirs have large volumes of carbon dioxide along with methane. One that I worked on was the Moore-Hooper Ellenburger gas field in Loving County, Texas. This field averaged about 50% methane and 50% carbon dioxide. However, the wells completed lower in the reservoir had up to 58% carbon dioxide, and the wells completed on top of the structure had only 42% carbon dioxide. This suggests the source of the carbon dioxide was from below, which would signify de-gassing of the planet.

The carbon dioxide comes from deep in the planet and has basically the same source as the carbon dioxide in volcanoes. However, we get carbon dioxide coming up deep fracture systems without any surface evidence of volcanism.

The volcanic gases believed to be closest in composition to those that have been given off throughout the earth's history have been found at Kilauea Volcano in Hawaii. These gases are from a deep oceanic source over a hot spot and probably came directly from the mantle of the earth. They were collected from the Halemaumau Lava Lake in 1919 and are shown in Table 3.

Table 3. Average Composition in Volume Percent (at 1,200 degrees Centigrade):

Water vapor	70.75%
Carbon dioxide	14.07%
Carbon monoxide	0.40%
Hydrogen	0.33%
Nitrogen	5.45%
Argon	0.18%
Sulfur dioxide	6.40%

Sulfur	0.10%
Sulfur trioxide	1.92%
Chlorine	0.05%

Table 3. Composition of Gas from Kilauea Volcano. This gas composition is the average from fourteen samples collected from Halemaumau Lava Lake in 1919. From *Volcanoes of the National Parks in Hawaii*, 1982 edition.

The largest component of the Kilauea gases is water vapor at 70.75%. Part of this may have been from rain or surface water instead of all from the volcanic source. However, most of the water from the volcanic gas would fall as rain. Only those gases remaining in the atmosphere would affect our evaluation.

Some of the volcanic gases would react with the surface of the earth. These would include sulfur dioxide, 6.4%, and sulfur trioxide, 1.92%, which combined with water would form sulfuric and sulfurous acid. In addition, hydrogen, 0.33%, and chlorine at 0.05% would combine with water forming hydrochloric acid. These three acids all react vigorously to cause weathering of the rocks on the earth's surface. The gas samples were collected at 1,200 degrees and the 0.1% elemental sulfur would fall out of the gas as solid sulfur upon cooling. This leaves in the atmosphere carbon dioxide, carbon monoxide, nitrogen, and argon.

The approximate composition of the atmosphere 4,000,000,000 years ago was 70% carbon dioxide, 27% nitrogen, 2% carbon monoxide, and 1% argon. This initial composition is based on several assumptions. First, the atmosphere came mainly from volcanic gases with the removal of water and those gases that react vigorously with the earth's surface for weathering. Second, the volcanic gases over the last 4,000,000,000 years were similar to the Kilauea volcanic gases. Third, the contribution of gases from space to the current atmosphere was minor.

EVOLUTION OF THE ATMOSPHERE

The atmosphere has continued to form, change, and evolve since it first appeared. Now let us look at the evolution of the atmosphere from 4,000,000,000 years ago to the current composition. The present composition of air is 78% nitrogen, 21% oxygen, 0.93% argon, 0.036% carbon dioxide, and trace elements. There are also variable concentrations of water vapor and particulate matter. It is important to keep in mind that volcanic gases have been added to the atmosphere and oceans continuously throughout the earth's history. However, volcanic replacement of carbon dioxide in the atmosphere may be diminishing over time. We do not know how long it takes to de-gas the planet. In addition, carbon dioxide and oxygen have continuously been removed from the oceans and atmosphere and have been deposited in rocks. The net effect is that the atmosphere has become thicker and denser over the last 4,000,000,000 years. Plants have had a major role in the evolution of the atmosphere and changes in the gas in solution in the oceans.

Plants take in carbon dioxide, water, and nutrients weathered from the earth and expel oxygen in their life processes. This is complex, but the net effect to the atmosphere and oceans is for plants to take in carbon dioxide and give off oxygen. This oxygen is the source of essentially all the oxygen in the atmosphere and oceans. It is also the source of the oxygen that has been removed from the atmosphere and oceans by combining with other elements. The red soils we see in many parts of the country are colored by iron oxide, and most of the oxygen in the iron oxide originally came from the atmosphere or oceans.

More than 3,500,000,000 years ago, plants first appeared in the oceans. They were very simple single-celled plants which left no fossil record. The oldest plant fossils that have been found are stromatolites, which are secretions of limestone. These fossil structures were formed by complex blue-green algae, and the oldest have been dated at 3,500,000,000 years old.

The carbon dioxide concentrations of the atmosphere have been declining because of this plant usage for the last 3,500,000,000 years. The carbon dioxide concentration 3,500,000,000 years ago was about 70% or 700,000 parts per million and today it is 360 parts per million. This is an average decline of about 200 parts per million every million years. The decline in carbon dioxide in the atmosphere may have slowed down when animal life first appeared in the oceans about 570,000,000 years ago. When plants and animals started on land, about 415,000,000 years ago, the decline in carbon dioxide in the atmosphere may have accelerated. This is because plants on land developed more rapidly than animals. Once animals populated the land, the decline in carbon dioxide in the atmosphere may have slowed down some. However, the direction is the same, toward less carbon dioxide in the atmosphere. This is because plants have a much greater biomass and are dominant in the relationship between plants and animals.

In the evolution of the other components of the ancient atmosphere, nitrogen is not very reactive and much that reaches the atmosphere from volcanism stays there. This accounts for the large nitrogen buildup in the atmosphere. Lightning causes some of the nitrogen to combine with oxygen to form nitrous oxide which is used by plants. In addition, nitrogen-fixing bacteria associated with some plants provide nitrogen from the atmosphere to plants in a form they can use. The carbon monoxide in the atmosphere combined with oxygen to form carbon dioxide when there was sufficient oxygen. Lightning provides the energy of ignition for this reaction. The argon in the atmosphere is inert and accumulated over time.

The shape of the atmosphere also has a major effect on life which led to extinction.

Shape of the Atmosphere

The atmosphere is layered into distinct units that have different chemical composition and density. The density of the atmosphere diminishes with altitude and in the high latitudes. For simplicity, we will look only at the troposphere, or lower layer which contains 85% of the atmosphere. The other layers, the stratosphere, ionosphere, and exosphere, have different compositions than the troposphere and have very low densities.

The troposphere is up to ten miles thick at the equator but only about five miles thick at the poles. The main reason for this is centrifugal force from the rotation of the earth which moves air from the polar regions toward the equator. The earth has a circumference of 24,903 miles at the equator and rotates once every twenty-four hours. This means that if we are at the equator we are traveling at a speed of 1037 miles per hour from the earth's rotation. Of course, we are not affected because we have gravity and the weight of the atmosphere holding us down. However, this centrifugal force deforms the atmosphere and pushes it out at the equator in a big way.

The approximate speeds that we would travel due to the rotation of the earth at various latitudes are as follows: 15 degrees — 1003 MPH; 30 degrees — 901 MPH; 45 degrees — 732 MPH; 60 degrees — 518 MPH: 75 degrees — 271 MPH; at the North and South Poles there is essentially no centrifugal force. Taking a closer look we see that there is only about 13% difference in centrifugal force from the equator to 30 degrees north and south latitude. This would suggest that there is only a minor difference in the thickness of the atmosphere from the equator to 30 degrees latitude. From 30 degrees to the poles, the magnitude of centrifugal force decreases greatly, indicating that the thickness of the atmosphere diminishes rapidly. The carbon dioxide available for plants would also diminish with the thickness of the atmosphere.

Centrifugal force is directed away from the earth's axis of rotation. The lines shown on the circumference of the earth are at 15 degree intervals, and show the direction of centrifugal force. The length of the lines represents the approximate speed from the rotation of the earth.

Figure 7. Diagram of Centrifugal Force.

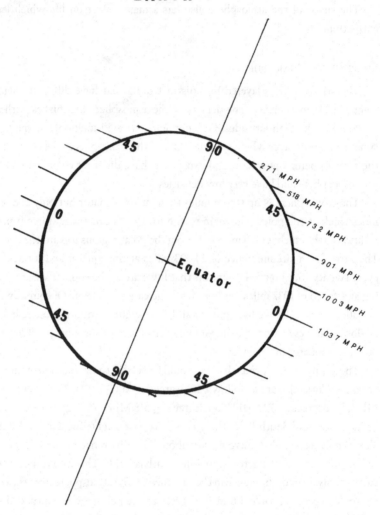

When we use a barometer to measure the air pressure, what we are actually measuring is approximately the gravitational force between the earth and the atmosphere minus the centrifugal force. The gravitational force on the atmosphere is directed toward the center of the earth. However, the centrifugal force is directed away from the axis of rotation so that the forces are not opposite except at the equator. The only place that a barometer actually measures the weight of the atmosphere is at the North and South Poles, where there is no centrifugal force.

A roller coaster that loops will more than offset the force of gravity by centrifugal force when riders are upside down at the top of the loop. In the same way, part of the force of gravity on the atmosphere is offset by centrifugal force from the rotation of the earth.

Centrifugal force also has an effect on carbon dioxide in the atmosphere. Since carbon dioxide is about 65% heavier than air, the earth's centrifugal force has a tendency to move carbon dioxide toward the equator. On a short term basis, this effect would be diminished by wind mixing. However, this process has been going on for millions of years and the cumulative effect should be significant. To measure the carbon dioxide in the air that plants can use, we should measure absolute carbon dioxide instead of relative carbon dioxide. We should measure carbon dioxide molecules per liter of air instead of parts per million carbon dioxide.

Since we cannot see the atmosphere, it may be difficult to visualize the effects of centrifugal force on the atmosphere. It may be easier to see the effect of centrifugal force from the rotation of the earth on the earth itself and the ocean waters. The distance from the center of the earth to the equator is thirteen miles longer than from the center of the earth to either Pole. The difference is caused by the centrifugal force from the rotation of the earth.

The radius of the earth is measured at sea level and the levels of the oceans are affected by centrifugal force. It may be easier to visualize centrifugal force on the oceans by looking at what happens if it is removed. If the rotation and centrifugal force were to stop, the oceans would flow from the equator to the Poles giving little ocean water at the equator and a water depth at the poles of about thirteen miles. Some of the atmosphere would also flow from the equator to the poles and it would be much thicker at the Poles.

Let us look at how the shape of the atmosphere affects life. The density of the atmosphere diminishes in the higher latitudes and with altitude. We can get a hint of how this diminishing density affects plant life by looking at the tree lines.

Tree Lines

The tree lines are lines on earth beyond which no trees grow. There are mountain tree lines, above which no trees grow, and Arctic tree lines or lines in the high latitudes beyond which no trees grow. In looking into tree lines and why trees stop growing, it is worthwhile to look at the requirements for tree growth.

A tree needs sufficient carbon dioxide from the air, water, and nutrients from the soil plus light energy from the sun for survival. If the ground is frozen to great depths in the winter and summer, then a tree cannot get water and nutrients from the soil and will not sprout and grow. If the growing season is too short to manufacture and store enough food to get through the dormant season, then the trees will not survive. If a tree cannot get sufficient carbon dioxide from the atmosphere, then it also will not sprout, grow, and survive.

A tree has thousands of leaves or needles with an extremely large surface area to take in carbon dioxide. This is because the concentration of carbon dioxide in the atmosphere is so low, at about 360 parts per million. Humans and other animals, on the other hand, have small nostrils to take in oxygen and a relatively small lung capacity to process it because the oxygen concentration in the atmosphere is 210,000 parts per million.

Higher in elevation and higher in latitude two things happen — the air gets thinner and the climate gets colder. The controlling factors on tree lines are availability of carbon dioxide and possibly temperature. The distribution of mountain tree lines indicates that the lack of sufficient carbon dioxide is the main controlling factor. Tree lines are the same elevation on both the sun side and the shade side of the mountain, suggesting temperature is not a limiting factor.

In addition to tree lines, there are different limits on all vegetation due to a lack of atmospheric carbon dioxide. These limits on all plant life are much fuzzier than tree lines because of the wider variety of small plants. Each type of plant has a different level of carbon dioxide needed to survive. In the mountains, the limit of most plant life is about 4,000 feet above the tree line.

Let us look at how elevation relates to the carbon dioxide of the air. When I worked in a mine at Climax, Colorado, we lived at about 10,500 feet in elevation, which was very near the tree line. The raw barometric pressure was about 19 inches of mercury at the tree line. Climax, Colorado is at latitude of about 39 degrees north. At sea level, the barometer reading is 29.92 inches of mercury. This is about a 36% reduction in air density as measured by a barometer. With a 36% reduction in current carbon dioxide, this is 360 parts per million times 0.64, which equals 230 parts per million. This means that if the carbon dioxide concentration in the atmosphere was 230 PPM, instead of 360 PPM, then theoretically the tree line at 39 degrees north (and south) latitude would be at sea level. In reality, winds, proximity to oceans, and other factors give very large distortions so that tree lines at sea level vary considerably, with 39 degrees latitude being merely the average.

Chapter 6. Changes In The Atmosphere During The Last Extinction

There are indications from studies of air inclusions in glacial ice cores that the carbon dioxide concentration of the atmosphere was about 180 parts per million in the last glacial breakup phase. This would put the tree lines around 30 degrees North and South latitude at sea level. Where elevations were higher, the tree lines and the limit of most vegetation would be closer still to the equator.

Current tree lines show sharp breaks in the trees with altitude. The trees grow very slowly and become short and stunted near the tree line. Trees grow up the mountains until they reach their survival level and stop. However, there are no sharp breaks in the high latitudes, which make Arctic tree lines from low atmospheric carbon dioxide much fuzzier. This is because surface winds move air of different carbon dioxide levels around in a horizontal direction.

In the last 150 years, the carbon dioxide in the atmosphere has increased from about 280 to 360 parts per million. There are two reasons for this. First, humans have put large volumes of carbon dioxide into the atmosphere through combustion and other activities. Second, the warming of the ocean's surface after the end of the Little Ice Age has reduced solubilities releasing carbon dioxide to the atmosphere. The reduced solubility makes it more difficult for the oceans to absorb more carbon dioxide. Because of the very slow growth of trees at tree line, the tree lines have probably not completely adjusted to this rapid change in carbon dioxide levels. We will now look at the change in the atmosphere in the active period of the glacial cycle.

LATE GLACIAL CYCLE ATMOSPHERIC CHANGES LEADING TO EXTINCTION

During the active phase of the glacial cycle, the temperature of the atmosphere rapidly increased. This was because the heat input to earth was rising and the surface temperatures of the oceans were rising in the early glacial spring. This increase in atmospheric temperature continued in most places throughout the active phase.

Precipitation increased early in the active phase because the surfaces of the oceans in the low and mid-latitudes were warming rapidly, increasing evaporation. Also with a warming atmosphere, the moisture evaporated from the oceans could be transported longer distances over the glacial sites, which increased snowfall on the glaciers. The high heat input to earth also produced melting of the glaciers and made them more active. Throughout the last half of the active phase, the melting and wastage exceeded the new snow accumulation, which cooled the ocean surface in the high latitudes. Consequently, the amount of carbon dioxide in the atmosphere fell throughout the last half of the active phase of the glacial cycle. After the active phase, there was a dramatic change in the atmosphere during the glacial breakup phase.

At the start of the glacial breakup phase, the heat input to earth was very high and reached its peak near the end. The breakup phase started when melting and wastage greatly exceeded new snowfall and glacial ice creation. As the meltwater and melting ice flows cooled the surface of the oceans, not much evaporation took place, resulting in less moisture in the atmosphere to re-supply the glaciers. This process rapidly moved to an essentially irreversible glacial breakup everywhere except Antarctica and Greenland. The surface of the oceans became the coldest in the cycle from the meltwater and floating ice melting. The precipitation dropped to the lowest point in the glacial cycle. There was also much less cloud cover to reflect energy back into space with the low moisture content in the atmosphere.

The temperature of the atmosphere had two very strong but opposite forces working on it. The highest heat input to earth in the glacial cycle melted most of the glaciers and heated up the continents and the atmosphere. The cold temperature of the oceans' surfaces had a tendency to cool down the atmosphere. However, there was not as much heat exchange between a cool ocean and a warm atmosphere as there would be between a warm ocean and a cool atmosphere. The net result was a warm, dry, generally cloud-free atmosphere over the continents and a cooler, relatively dry atmosphere over the oceans.

The cold ocean surface, composed mainly of meltwater runoff, melted icebergs, and ice flows, had a low content of atmospheric gases and a large capacity to absorb them. The volume of meltwater released from the glaciers during the last breakup phase could have held all the carbon dioxide in the atmosphere at zero degrees centigrade without being completely saturated.

The Atmosphere during the Interglacial Period

The interglacial period that we are currently in is the warmest of the glacial cycle. The top portions of the oceans started very cold from the glacial breakup and warmed rapidly. The temperature of the atmosphere also warmed. The atmosphere and the oceans continued the warming trend throughout the interglacial period.

As the top portion of the oceans warmed, the ocean water could not hold all the carbon dioxide that was absorbed during the glacial breakup phase, so some of it came out of solution. Carbon dioxide levels in the atmosphere increased throughout most of the interglacial period. With this increase in carbon dioxide, plants migrated to the higher elevations and higher latitudes. The animals that ate the plants followed their food source. Humans that ate both the plants and animals also followed their food source to the higher elevations and higher latitudes.

PARTICLES IN THE ATMOSPHERE

Other events cause particles in the air and alterations to climate. An example might be the blinding blizzards that hit eastern Kansas in the winter of 1885-1886. The following spring and early summer, it was so cold that the wheat did not mature and cattle that survived the winter had nothing to eat. I believe the reason for this very cold year was the cumulative cooling effect of large amounts of volcanic dust in the atmosphere for the prior two years. In August, 1883, halfway around the earth in the Sundra Strait between Sumatra and Java on the island of Krakatoa, there was an extremely violent volcanic eruption. This island all but disappeared — many cubic miles of island were blown away. A large amount of volcanic debris was thrown high into the atmosphere as volcanic dust, where it circled the earth. The volcanic particulate matter in the atmosphere reflected more of the incoming sunlight and energy back into space, causing a cooling of the atmosphere. This was a time of spectacular sunsets caused by the dust.

Today particulate matter in the atmosphere comes from many sources besides volcanoes. Utility plants and industry eject large amounts of particulate matter through the smoke stacks of the world. Although it is a problem in the United States, it is a much greater problem in many third world countries. The United States is prosperous enough to have developed more sophisticated industrial methods and can afford to implement relatively strong environmental protection laws. In many parts of the world, much of the industry and utilities are marginal and antiquated in design, and the problem of feeding the population is more urgent than finding a way to pay the West to bring in expensive systems to reduce particulate matter pollution.

A very large source of airborne particulate matter is soil erosion by wind. While the United States and many other countries have learned to do a better job of preventing the soil from blowing away from farm land, the other hand, there is a large increase in cultivation throughout the world. The expansion of land under cultivation often involves marginal terrain that is too dry or too poor for efficient farming and is more likely to be affected by wind erosion. Much of this marginal land is in less-developed areas of the world, where wind erosion is not well controlled. The net result is that wind erosion overall is becoming more severe. Its effect is putting greater quantities of particulate matter into the atmosphere.

In some countries, there is a considerable amount of burning associated with agriculture and the lumber industry, which contributes significant particulate matter into the atmosphere. During the major fires in Yellowstone National Park in 1988, a significant smoke haze occurred over Dallas, Texas, which is 1,300 miles away. The wind was out of the northwest and brought the smoke from the Yellowstone fires over the city. This is an indication of how long smoke can remain in the atmosphere.

The particles from industrial sources, burning, and wind erosion of the soil all act much the same as volcanic dust. These particles usually do not go as high in the sky, which means they drop out of the atmosphere sooner. They still reflect more of the sunlight and energy back into space and have a cooling effect on the earth.

Anyone can make a comparative estimate of the particulate matter in the air by observing the length and intensity of twilight. Twilight is caused by the sun below the horizon hitting particles higher in the atmosphere and scattering the light. The longer and more intense the light is after sundown and before sunrise, the greater the particulate matter in the atmosphere.

In the early 1970s, I lived in Calgary, Alberta, and the January sunrises were amazing. It appeared dark until the sun reached the horizon, and within two minutes, it was fully light. There was only a minor twilight. The reason was that snow cover had been on the ground for two months, keeping the dust on the ground. What little wind there was came from the northwest, where there was no industry to put any particles into the air. In addition, because of the extreme cold, there was very little moisture in the air.

By comparison, a normal March Kansas twilight lasts about an hour and is quite bright. In Kansas, almost all of the particulate matter is windblown dust from soil erosion. In the polar regions, by contrast, there is a very minor but discernible twilight that lasts all day every day for about a month. Here the low intensity is due to very few particles in the atmosphere, and the long duration is caused by the sun being just below the horizon for a whole month.

This wide variation in the distribution of particulate matter is one of the reasons that it is not taken into consideration as a factor for altering the temperature of the atmosphere. It is very difficult to quantify, but we know from Krakatoa that it has a significant effect.

Chapter 7. Glacial Changes to the Land Affecting Life

To see the effect of glacial changes on plants, animals, and humans leading to the large extinction, we need to look first at some of the things that glaciers do to the land. Glaciers dramatically change the land. They do far more physical damage to the land and the habitats of plants, animals, and humans than people and their machines can ever do. Glaciation, and the associated rise and fall of sea level, causes major changes on all land surfaces. There are some glacial effects today because about 10% of the land surface is covered by glaciers. However, most of what we are talking about is land changes during the accumulation and dissipation of continental glaciers. At the peak of the last glacial cycle, it is estimated that 27% of the present land surface was covered by glaciers.

The changes to the land caused by the glacial cycle are both directly caused by the glaciers and indirectly caused by glaciation. All of these changes affect plants and animals, and many affect humans. The direct land changes are caused by the formation, growth, and movement of glaciers, and by the water that comes from the glaciers when they break up, melt, and dissipate. The indirect land changes of the glacial cycle occur in areas not touched by glaciation and are caused by the rise and fall of sea level and by atmospheric changes related to the glacial cycle. No two glacial cycles are alike, so that the glaciation does not always cover the same area.

What Glaciers Do to the Land

Glaciers directly cause major changes to the land surface as they form, grow, and move across the land. Even the mass of the continental glaciers causes changes in the land surface. The land surface is depressed by the weight of the ice as glaciers build over the continental areas. In the last glacial cycle, very thick glacial ice over Hudson Bay in Canada depressed the land surface by over a thousand feet. This is a very slow process and it takes a long time for the land surface to be depressed by the weight of the ice. It also takes a long time for the land surface to rise back to its original level after the ice has melted. Hudson Bay and the Great Lakes are still rising today in response to the glacial ice that melted 12,000 years ago. These up-and-down movements of the land's surface from loading and unloading ice are called isostatic readjustments.

A glacier is composed of ice, rock, and water. As it increases in size and starts to move outward, it becomes a very strong force in changing the surface of the land. Gravity moves the glacier with the rocks and boulders that the glacier picks up. These rock components provide cutting tools to reshape the land. The continental glacier both carves rock and soil from the land in some places and deposits rock and soil in other places. The glacier is tough on all living things, plowing down trees and destroying all vegetation in its path.

If the surface of the land on which the glacier forms is fairly flat, then the glacier slowly moves out in all directions as the glacial mass builds. Continental glaciers can move out for many hundreds of miles in each direction and grow to many thousands of feet thick.

When continental glaciers form in mountainous regions, they create much more dramatic changes than on the flat land. The vertical drop in the mountains is much greater so that erosion from the force of gravity is much greater. The V-shaped valleys formed by water erosion of rivers are cut into U-shaped valleys by glaciation. In the large U-shaped valleys, glaciers are also supplied with ice from side valleys flowing into them. Often, these side valleys are much higher than the U-shaped valley floor and are called hanging valleys after the ice is gone. These are the sites of some of our most spectacular waterfalls today. Yosemite Falls, with its total drop of 2,425 feet, is an example of a waterfall coming from a hanging valley.

As a mountain glacier starts to move, it plucks rock from the surface, forming a basin in the mountains called a cirque. When several glaciers grow together, they sculpt all sorts of intricate features. They often leave very sharp high peaks and spires, called horns. Two famous horns are the Matterhorn in the Alps and Mount Assinibone in the Canadian Rockies. Sometimes, the combination of cirques and horns produces features called serrate ridges. An example of a serrate is at the Valley of Ten Peaks surrounding Moraine Lake in Banff National Park in Alberta, Canada.

Chapter 7. Glacial Changes To The Land Affecting Life

When continental glaciers and sea ice form, there is a drastic change in albedo or reflectivity of the earth's surface. When the albedo is small, more of the sun's energy is absorbed on the surface of the earth. The land surface before glaciation has a small albedo. After glaciation, the new snow surface reflects 90% of the incoming light energy, and old snow reflects 50%. When continental glaciers are in place, it takes a very large amount of energy to overcome the albedo and melt the glaciers.

In addition to albedo and how the incoming energy is utilized, there are large changes in glaciation caused by changes in the energy input from the sun. At the end of the active phase, the energy input to earth was approaching the highest in the cycle. This was when the glacial breakup started. The direct changes to the land from growing and moving glaciers diminished fairly fast. These changes were replaced by the alterations to the land from the glacial breakup and the vast amounts of meltwater flowing to the oceans. Next, we will look at how the glacial breakup affected the land.

Land Changes from Glacial Breakup and Melting

The last glacial breakup phase may have been shorter and different from previous breakup phases. Since it is the most recent major glacial event, there is a lot of field evidence that has been described and interpreted over the last 200 years.

The heat input from the sun during the last glacial breakup phase was close to the highest heat input in the cycle. As the glaciers melted, vast amounts of very cold water flowed away from the glaciers. When the glaciers moved, they scoured the land, picking up large amounts of soil, rocks, and even massive boulders. Most of this sediment had been incorporated in the glacier. The meltwater that flowed away from the dissipating glaciers carried large quantities of this sediment from the glaciers. Large glacial till plains composed of glacial boulders, rock, soil, and ground-up rock called glacial flour were deposited directly in front of the glacier. Glacial till has the character of its source. Many pebbles came from a granite source, sand came from a sandstone source, clay tills came from a shale source, and rock flour came from a limestone or dolomite source. As the glaciers receded, glacial till plains were deposited where the glacier previously stood. Farther away from the glacial front, streams flowing from the glacier deposited glacial outwash and formed outwash plains. Glacial outwash has a smaller particle size than till and is better sorted.

The water from the glaciers formed a great number of lakes and some of these were of massive size. There was a large volume of fine sediment deposited in these lakes. One of these very large lakes was Lake Bonneville in Utah. A small remnant of this massive lake is the present day Great Salt Lake. Most of the land speed records by automobiles have been established running on the flat, dry, ancient lake bed of Lake Bonneville.

Much of the glacial meltwater moved to the sea continuously in very large channels such as the Mississippi River. It also moved to the sea in great episodic events when extremely large lakes broke their dams and drained rapidly. These dams may have been ice dams or sediment dams. These large draining events caused massive and rapid erosion. The Grand Coulee, which is a very large channel in Washington State, was formed by these episodic events. The Saint Lawrence Seaway had continuous drainage in parts of the seaway and episodic draining events from very large lakes formed on the upper channel. After the glaciers dissipated there were many glacial features remaining on the land.

Land Changes Left Behind After Glaciation

Glaciation wrought many changes to the land which are apparent after the glacial breakup. Some of the evidence is clearly visible today. Everywhere that large volumes of ice existed, the surface of the earth was depressed by the weight of the ice. When the ice melted and flowed into the oceans, this land started to rise for an isostatic readjustment. These movements of the earth's crust are very slow. We can still see remnants of some of these depressions. Where the ice was thickest this upward readjustment is still going on. Areas including parts of the North Sea, Baltic Sea, Hudson Bay, and the Great Lakes are still rising and some day parts of these may be dry land.

When the glaciers melted, the rocks that were in the glaciers were left on the land. At the old glacial front large piles of glacial till, called terminal moraines, were deposited. Much of Long Island, New York, is an example of a large terminal moraine. In mountain areas there are also lateral moraines where rocks from along the valley walls fell onto the glacier and were incorporated into its sides. When two mountain glaciers flowed together, the lateral moraine from one side of each glacier ended up in the middle of the combined glacier and is called a medial moraine. When the glaciers melted, the moraines were left on the land as large ridges.

Erratics are boulders that were moved a long distance by the glaciers and are composed of very different types of rock than those commonly found where they were deposited. Drumlins were left by retreating glaciers and are smooth elongated hills averaging about 100 feet high and a half mile long, made up of glacial till. Many drumlins may be found grouped together in drumlin fields. Eskers are long winding ridges that the glaciers leave behind. They can be ten to fifty feet high and a quarter mile to eighty miles long.

There are many glacier related features left by the rivers and streams flowing from the ice. Many lake beds were carved out by the moving ice and rocks of the glaciers. As the ice melted, these cut-down areas filled with water to form lakes. Some large blocks of ice were buried, and as they melted, the overlying ground collapsed and formed potholes which filled with water.

Glaciation also caused extensive indirect changes to the land in non-glaciated areas around the globe. Next, we will look at indirect glacial effects that created large loess and sand dune deposits at the end of the last glacial cycle.

Events Leading to Loess and Sand Dune Deposits

There is a long and convoluted sequence of events leading to the formation of loess and sand dune deposits at the end of the last glacial cycle. The last glacial breakup phase was caused by the heat energy From the sun being near the highest in the glacial cycle. The melting glaciers shed extremely large quantities of very cold meltwater into the oceans cooling the surface. Sea ice in the oceans surrounding the continental glaciers floated away and melted. The glaciers that terminated in the oceans calved large icebergs which melted slowly as they drifted toward the equator. As sea level rose, glacial ice that was grounded on the sea floor was floated off the bottom and drifted toward the equator and slowly melted. It takes one calorie of heat to raise the temperature of one gram of water one degree Centigrade. However, it takes 80 calories of heat to melt one gram of ice. Consequently, the melting of large quantities of ice near the surface of the oceans had a major cooling effect on the surface of the oceans.

This was a time of high heat energy input to earth, but much of the heat reaching the oceans did not heat up the surface. As we noted before, 92% of the energy of the sun reaching the ocean is absorbed, but not at the surface. Half of the energy is absorbed in the top thirty-three feet and the other half is absorbed down to the depth of sunlight penetration. This means that a lot of energy can be absorbed by the ocean without a major surface warm up.

In the last glacial breakup phase, all of this very cold meltwater flowing to the ocean and ice melting in the ocean was adding fresh water to the oceans,. This cold fresh water was about 3% lighter than the sea water and most of it floated on top of the salty sea water. This cold fresh water was undersaturated in atmospheric gases. As the surface of the oceans cooled, then more carbon dioxide went into solution in the oceans and was removed from the atmosphere. The carbon dioxide levels in the atmosphere reached the lowest point in the glacial cycle near the end of the breakup phase. The atmospheric carbon dioxide levels at that time were half of the current levels. This affected the land in a very dramatic way where the plants died off from lack of carbon dioxide, and high winds eroded the barren lands and glacial sediments and deposited large quantities of wind blown particles.

At the same time with the surface of the oceans very cold, not much water vapor was evaporated from the oceans. The high heat input to the earth from the sun made the land surfaces and the atmosphere over the land surfaces very warm. Warm air can carry many times more water vapor than cold air. Consequently, air over the land had a high capacity to carry moisture and there was very little moisture evaporated from the oceans to be carried. The result was hot dry continents with very little moisture falling anywhere, except in some areas of the tropics. We will now see how low atmospheric carbon dioxide and low precipitation leads to loess and sand dune deposits.

Origin of Loess and Sand Dune Deposits

With falling carbon dioxide levels in the atmosphere, many plants could not survive. The tree lines and the limits of most vegetation migrated to lower latitudes and lower elevations, creating barren land during the breakup phase. The low precipitation and ensuing droughts killed many plants. As the glaciers melted and receded, they left behind glacial till, glacial outwash, and large dry lake beds. These deposits were composed of large volumes of loose, easily eroded silt and sand-sized particles that were very susceptible to wind erosion.

The results of these factors were large areas of barren land devoid of vegetation in the higher altitudes and mid-to-high latitudes. In addition, there were treeless lands with greatly diminished vegetation covering extensive areas. The atmosphere was relatively cold over the remaining glaciers, hot over the continental areas, and cool over the oceans. These temperature extremes caused large pressure differences in the atmosphere, resulting in high winds over the continents.

When there are high winds over dry, barren lands composed of loose particles, dust storms are a big part of the daily weather patterns. Large loess and sand dune deposits were formed over much of the Great Plains. Loess is wind-blown dust composed of silt sized particles. Some loess deposits are associated directly with glaciation, being composed mainly of particles left by glaciation. Other loess deposits are formed by wind-eroding soil particles from barren ground or ground with sparse vegetation not associated with glaciation.

There are many, very large loess deposits associated with the breakup phase of the last glacial cycle. They are located between 30 and 70 degrees north latitude in North America, Europe, North Africa, and Asia, and from 25 to 40 degrees south latitude in South America and Australia. These deposits cover much of the plains in the areas shown on the loess and sand distribution map in Chapter 3. These loess deposits must have been extremely large when they were formed. Wind and water erosion have been removing sediment from these easily eroded deposits for the last 12,000 years and they are still significant. Much of the soil in the plains of the United States is loess that was formed 26,000 to 11,000 years ago.

One of the most striking examples of a loess deposit is found at Loess Hills, Iowa. These are a band of hills running north to south for 200 miles on the west side of Iowa. They are formed directly east of the Missouri River flood plain. They are made up of three deposits of loess. All of these deposits were formed over 12,500 years ago. The Loess Hill deposits have about 200 feet of loess at their thickest point. They are believed to be deposited by prevailing westerly winds.

Even some ice cores from Greenland have high dust concentrations deposited during the time of the glacial breakup phase. This indicates that the dust storms were very large, and high wind carried the dust a long distance. Where the dust was deposited on the glaciers, there was a significant reduction in the albedo of the glaciers. They would absorb much more energy from the sun and melt much faster than they would have without the dust on the glaciers.

In addition to the larger loess deposits formed from the high winds over barren ground, there were also deposits of larger particles in sand dunes. These were also formed by high winds over barren land. One of these sand dune deposits is found in Colorado.

The San Louis Valley in south central Colorado has an elevation between 7,000 and 8,500 feet. The valley floor is composed mainly of sand eroded from the San Juan Mountains to the west and the Sangre de Cristo Mountains to the east. During the breakup phase, the San Louis Valley is believed to have been barren. The prevailing wind is from the west. This combination of barren sandy ground and a prevailing west wind provided the conditions capable of forming great sand dunes on the east side of the valley up against the mountains. Strangely enough, this just happens to be the location of the Great Sand Dunes National Park. These sand dunes cover fifty square miles and rise to a height of 800 feet above the valley floor.

In western Nebraska, there is an area of sand dunes formed by high winds over the barren lands of the plains during the glacial breakup phase. They also have been reworked later at various times during the interglacial period. These are not as spectacular as the Great Sand Dunes because they are relatively low and stabilized by vegetation. However, they are much more extensive and significant as they cover about 20,000 square miles.

Sand storms have a lot of power and wind-blown sand is quite an effective abrasive. In one afternoon it can pit the paint on a car, or strip off patches of paint. Over a longer period of time, wind-blown sand cuts into rocks on the surface of the land, resulting in more sand and many smaller dust particles that become loess.

The very large loess and sand dune deposits of the glacial breakup phase are a result of low carbon dioxide and low precipitation. The 6,000 years of the glacial breakup phase saw the most rapid change in the land surface. Changes in sea level cause other indirect alterations to the land surface.

Land Changes from Lowering Sea Level

As water is evaporated from the oceans and is deposited as snow and ice on land to form glaciers, the sea level drops. Throughout the 40,000 years of the accumulation phase and early part of the active phase of the last glacial cycle, there was a 350 foot lowering of sea level. Many alterations to the land surface occurred because of this drop in sea level.

When the sea level is lowered, there is an increase in gradient or slope on all rivers and streams draining into the oceans. The steeper gradient means that the river runs faster, and the river bed is cut down faster by water erosion. With enough time, a stream will cut down the stream bed all the way back to the headwaters of the stream. Stream beds that are relatively flat in soft sediments erode a deep channel. Streams that flow from the mountains into the sea already have a steep gradient, so the lowering of sea level causes less change in down-cutting capacity.

On relatively flat rivers like the Mississippi, Amazon, and Niger, the lowering of sea level has a great effect. These rivers cut deep valleys in the land because of the steeper slope to the oceans. The area where New Orleans is located today was a valley floor over 300 feet lower when glaciation was greatest and sea level was the lowest. There was also faster erosion throughout the land around these rivers because of the lower sea level and lower river valleys. The earth's rivers moved larger amounts of sediment to the oceans because of the increase gradient.

With the lowering of sea level throughout the many glacial cycles of the Pleistocene there has been repeated down cutting. The rising sea level during the glacial breakup phases caused drastic changes.

Land Changes from Rising Sea Level

During the glacial breakup phase of the last glacial cycle, the oceans rose about 300 feet in approximately 6,000 years. This gave an average rise of one foot every twenty years. From a geological point of view, this is extremely rapid. The surface of the land was also altered rapidly from this rise in sea level.

With the melting of the glaciers, large volumes of water and sediment were transported toward the sea in river and stream channels. As the oceans rose, the rivers and streams slowed down with the flatter gradient. The larger sediment particles fell out of the slowing rivers and started to fill up the large deep river valleys that were formed when the ocean level dropped. This process of sediment deposition building up the river beds in slowing rivers is called aggrading. The aggrading to fill these valleys was rapid because there was a lot of loose sediment associated with the glaciers. However, very few of the valleys were filled in the 6,000 years of glacial breakup.

At the end of the glacial breakup phase about 12,000 years ago, the Mississippi River Valley may not have been filled with sediment. New Orleans is on the Mississippi Delta today, and then it was probably on the ocean floor. More valley-filling and the creation of deltas, such as the Mississippi and Nile deltas, occurred in the interglacial period. Today the Columbia River and Saint Lawrence River are still aggrading. Some of the valleys had a very small sediment source and are now mainly filled with sea water. Examples of these are the fjords of Alaska and Scandinavia.

The rising of sea level by 350 feet drowned the continental shelves throughout the earth. Many shallow seas returned on very low land that had been exposed for over 50,000 years. Many land bridges were once again covered with water. This changed many circulation patterns of the oceans. This also altered or stopped some migration of plants, animals, and humans. Next, we will look at the many changes in plants and plant distribution leading to extinction.

Chapter 8. Changes in Plants Leading to Extinction

Plants need carbon dioxide from the air, energy from the sun, water and nutrients from the soil, and room to grow in order to survive. Plants also need a growing season long enough to make sufficient energy to last through the dormant season. During the glacial cycle, the availability of these basic requirements change, which causes major changes in plants and plant distribution. Obviously, in an area that had lush vegetation but is now covered with a mile thick glacier, the plants are gone and the plant distribution is drastically changed. Some of the changes in plants and plant distribution were not so obvious.

How do we know which plants lived in the glacial area before the glaciers came, and where they lived? In most cases, little or no direct evidence remains after the glaciers swept everything away. Some plant material that the glaciers over ran may have ended up in the terminal moraines left on the land after glaciation, but there is no way to identify where these plant remains came from. Land plant fossils are not normally very abundant or well preserved. Nature has many efficient methods of recycling plant material for reuse. The plants must have rapid burial in a reducing environment in order to be preserved. Termites, other insects, bacteria, and fungi make short work of most remains of vegetation. Pollen is often preserved in lake sediments, which enables us to form an idea of what plants were in the area of the lake. However, since pollen is wind blown, the location of the original plants is often questionable.

If no plant fossils are found, either no plants were present or none were preserved. Many plant species living today have still not been discovered or cataloged. Even so, it is suggested that there was a significant loss of plant species at the end of the last glacial cycle that has not been recognized. The plant loss would have been mainly in plants that had higher atmospheric carbon dioxide requirements. To put this into perspective, it is worthwhile to look at how plants have changed throughout geological history.

Plant Changes throughout Geological History

Plants originated in the oceans more than 3,500,000,000 years ago. The first plants were single-celled organisms that left no fossils. It probably took millions of years for plants to evolve enough to arrive at a plant that left fossil structures. The first plant fossils, called stromatolites, were formed by blue-green algae. The oldest stromatolite fossils are 3,500,000,000 years old. The blue-green algae are fairly complex lime-secreting plants which still exist today.

As noted previously, land plants take carbon dioxide from the air and give off oxygen. Animals take oxygen from the air and give off carbon dioxide. Plants have always been dominant in this relationship. The carbon dioxide content of the atmosphere has dropped about 200 parts per million every million years for the last 3,500,000,000 years. Consequently, land plants have had to evolve continuously throughout geological history to survive on less and less atmospheric carbon dioxide.

Simple plants first appeared on land in the Silurian period of the Paleozoic Era about 435,000,000 years ago. These plants had horizontal stems and upright branches and are called psilophytes. The atmosphere in the Silurian is believed to have had about 80,000 parts per million carbon dioxide as opposed to 360 PPM today. Consequently, it was much easier for plants to take carbon dioxide out of the atmosphere and there are indications that they grew very fast.

Plant evolution continued through the Devonian period, and by early in the Mississippian period, about 360,000,000 years ago, much of the land surface was covered with forests. They were composed of giant trees such as horsetail, seed fern, and club moss. The only horsetail and club moss living today are small herbs. These forests lasted until the end of the Pennsylvanian period, about 290,000,000 years ago. Many of the major coal deposits of the world were formed from these forests. The ancient forests grew in low lying swamps in an environment that was capable of preserving the hydrocarbons; over time, heat and pressure combined forces to form the plant debris into coal. While carbon from some of the carbon dioxide in the atmosphere used by the plants was stored in the coal beds, the oxygen from this carbon dioxide was released to the atmosphere.

European stratigraphers lump the Mississippian and Pennsylvanian periods together and call it Carboniferous because of the large coal deposits in the continental sediments and the large oil and natural gas deposits in the marine sediments. Most sediment from the Pennsylvanian period has a high carbon content in addition to the carbon content of the coal, oil, and natural gas generated in these sediments.

At the start of the Mesozoic Era (about 240,000,000 years ago), the carbon dioxide content of the atmosphere was probably about 50,000 parts per million. Throughout the Mesozoic, plants took carbon dioxide out of the atmosphere and released oxygen to the atmosphere. Some of the carbon was stored in the huge Mesozoic oil deposits, such as those in the Middle East and the tar sands of Canada. During the Mesozoic, gymnosperms (meaning "naked seed") were the most plentiful plants. The gymnosperms include conifers, cycads, and ginkgoes. The conifers have needles or scale-like leaves and include pines, firs, redwoods, spruces, cedars, and cypress. Cycads have trunks without branches and have long leaves at the top like a palm tree. However, cycads bear their seeds in large cones. Only one type of ginkgo survives today, and it is a tree with flat fan-shaped leaves. It is a native of Asia and is sometimes called the Maidenhair tree.

The flowering plants called angiosperms became dominant during the Cenozoic era, starting about 65,000,000 years ago. By that time, it is estimated that the carbon dioxide content of the atmosphere had diminished to about 13,000 parts per million. The flowering plants had a design that could more effectively utilize the carbon dioxide available. Some of the first flowering plants were trees, such as maples, magnolias, and oak trees. Those plants have evolved throughout the Cenozoic era.

There is a different atmospheric carbon dioxide requirement for each plant. Today plants have developed strategies to survive and thrive with atmospheric carbon dioxide concentrations at less than 360 parts per million at sea level in fairly high latitudes. Evolutionary changes to adapt to low atmospheric carbon dioxide would include developing more leaf area as a percentage of the total plant to take more carbon dioxide out of the air. Another strategy is to grow at a much slower rate. We are very close to the lower atmospheric carbon dioxide limit where all plants can survive. We have undoubtedly passed the lower limit for most land plants throughout geological history to survive.

Plant Changes in the Last Glacial Cycle

The conditions at the start of the accumulation phase of the last glacial cycle were fairly good for plant growth in many parts of the world. There was far more precipitation than the historical average throughout the world and it brought ice and snow to only the high latitudes and high elevations, with rain in most areas.

Carbon dioxide in the atmosphere was near the highest level in the glacial cycle, which was very advantageous to plant growth. The oceans and atmosphere were fairly warm, which was to the plants' liking.

As the accumulation phase continued, the precipitation increased and more of the land was covered by glaciers. The glaciers were fatal to the plants that had lived there previously. In the last half of the accumulation phase, the conditions for plant growth deteriorated. The atmosphere was cooling and carbon dioxide levels were dropping.

Near the end of the accumulation phase, the large continental ice sheets were essentially in place and the vegetation that had previously occupied the land was destroyed. The temperatures of the oceans and atmosphere were dropping rapidly. The carbon dioxide level in the atmosphere was also dropping rapidly. Periglacial conditions were expanding around the ice sheets which greatly restricted the growth of vegetation in these areas. However, the advance of the glaciers was slow enough so that most plant species were able to migrate away from the glaciers and relatively few plants became extinct during this time. Each year some of the seeds of the plant communities were blown to areas of more favorable climate, so that over time the plant communities migrated ahead of the glacial advance. As the glaciers covered some plants, new plants grew farther away.

The level of the oceans dropped about 350 feet by the end of the accumulation phase. Vegetation migrated to the newly exposed continental shelves and thrived on most of them, except in the higher latitudes. The building of the continental glaciers greatly restricted the areas available for growth of vegetation during the accumulation phase. Vegetation close to the new glaciers diminished because of the change in climate and lower atmospheric carbon dioxide. The survival potential for plant life continued to decline in the dormant phase.

The conditions for plant growth were poor at the start of the dormant phase and they deteriorated throughout the first half of the dormant phase. The oceans and atmosphere were very cold. Consequently, there was very little water evaporating from the oceans, and cold air cannot carry much moisture far over land. In addition, with the cold oceans, carbon dioxide was removed from the atmosphere and it went into solution in the oceans.

In the last half of the dormant phase, the oceans and atmosphere began to warm. This slowly increased the carbon dioxide in the atmosphere and increased the precipitation. All of these things started to bring the plants out of the deep freeze of the glacial winter. This improvement in conditions continued into the glacial spring and the active phase.

Chapter 8. Changes In Plants Leading To Extinction

The environment for plant growth improved early in the active phase. The temperatures of the oceans and atmosphere were rising. This released some carbon dioxide from the oceans. In addition, precipitation increased throughout the active phase. The combination of increased atmospheric carbon dioxide and adequate rainfall made for increased plant growth. Plant life flourished and migrated into some barren lands of the dormant phase which were now capable of supporting plant life.

Plant Changes Leading to Extinction

Late in the active phase, cold meltwater on the surface of the polar oceans, which was undersaturated in atmospheric gases, was removing carbon dioxide from the atmosphere. Lower atmospheric carbon dioxide started to have a detrimental effect on marginal plants, such as those in Australia.

In the northern hemisphere, about 18,000 years ago, the glacial breakup started with severe consequences for most land plant communities. The surface of the oceans over most of the earth was cooled dramatically with the influx of meltwater and floating ice as the glaciers broke up. The cold ocean waters took more carbon dioxide out of the atmosphere into solution. In addition, not much water evaporated from the oceans because the surface was so cold. This gave rise to cool dry air over most of the oceans, while over most of the continental areas the atmosphere was hot and dry. Consequently, two of the requirements for plant growth (atmospheric carbon dioxide and precipitation) were in short supply throughout most of the planet.

Near the end of the glacial breakup phase, the atmospheric carbon dioxide levels reached their lowest point in the glacial cycle at about half the current levels. This was also the lowest atmospheric carbon dioxide level since plants had appeared on earth. The very low atmospheric carbon dioxide levels probably lasted from about 16,000 to 11,000 years ago.

At the lowest point, the average tree line at sea level was believed to be below 30 degrees north and south latitude. In the United States, this would be at Houston, New Orleans, and Jacksonville. In areas of higher elevation, the tree lines were much closer to the equator. The average limit of most vegetation at sea level was believed to be below 40 degrees north and south latitude. This would be near Philadelphia and Madrid, Spain. The limit of most vegetation was much closer to the equator in the higher altitudes.

Carbon dioxide-rich ocean currents gave off carbon dioxide as they warmed up. This distorted tree lines and the limit of most vegetation on land adjacent to the oceans. Prevailing winds often moved air with higher carbon dioxide levels to other areas, which altered the tree lines and limits of most vegetation.

In the eastern United States, there were boreal woodlands throughout most of the last glacial cycle, and most died out during the period of low atmospheric carbon dioxide. After the atmospheric carbon dioxide increased, a mixed coniferous-deciduous forest migrated into the eastern United States. The boreal woodlands migrated into previously glaciated areas of Canada. Vegetational changes following a similar pattern also occurred in Europe at this time.

The low atmospheric carbon dioxide levels also affected the areas near the equator. In mountain areas, the tree lines of the tropical rain forests were at elevations approximately 7,000 feet lower than they are today. Until recently, researchers have considered that lower temperature, rather than lower atmospheric carbon dioxide levels, was the factor limiting the growth of the tropical rain forest. In addition to the tree lines, it follows that the limit of most vegetation in the tropics would also be approximately 7,000 feet lower than today. However, this would only be applicable on very tall mountains, such as Kilimanjaro in Africa and the Andes in South America.

In our previous discussion of the barren lands in the breakup phase, we identified conditions for the middle latitudes which created very high winds with dust and sand storms. Let's examine what these conditions did to the plants next to the barren lands by looking at a recent example.

In the early 1950s, there was a dust bowl in the Great Plains states. This was essentially reworking 14,000-year-old loess deposits. It was not as spectacular and debilitating as the dust bowl of the "Dirty Thirties," but it covered a larger area, extending father south into Texas and farther north into Colorado and Nebraska. However, in some places the wind erosion and dust in the air was quite severe. The way the wind damage started and spread in a wheat field is probably similar to what happened on a much larger scale during the breakup phase.

During the drought, the land underlying western Kansas wheat fields dried out and this put a great deal of stress on the young short wheat. The dust started being blown around so that soon there were small areas where the wheat was covered with dust, causing it to smother and die. This loose dust was then blown to adjacent areas, killing more wheat. The bare areas were subject to more rapid erosion. In a few weeks, the whole wheat field was dead and large amounts of blow dirt (loess) from one field moved to adjacent fields, killing them. This process occurred in many places in the Plains states, causing extensive damage to the plants and the land.

In the 1950s, most of the dust storms occurred in the winter and early spring. However, during the breakup phase of the last glacial cycle, the temperature dynamics of the atmosphere probably made high winds blow more or less throughout the year. The high altitude west-to-east jet stream that is in place today was also at work then and it moved high-flying dust to the east. In the winter and spring in the northern hemisphere, the surface winds were mainly from the north-northwest to the south-southeast. This carried much of the loess and sand from the barren lands onto the land between the limit of most vegetation and the tree line. Most of the vegetation on this land was probably similar to the Arctic tundra or short grass prairies of today, although stunted by low carbon dioxide and drought. Much of this vegetation was destroyed by the advancing loess and migrating sand dunes.

In the summer and fall in the northern hemisphere, the prevailing winds were probably blowing in a south-southwest to north-northeast direction. This brought loess and sand back from the smothered tundra and grasslands to the barren lands. The repeated dust and sand storms, shifting the loess and sand to the south and then to the north every year for thousands of years, destroyed most of the vegetation from the tree line north. This also did major damage to the animal life that lived there.

A similar process was occurring in the southern hemisphere. During the last glacial breakup, there was some barren land in southern South America and the adjacent exposed continental shelves. North of the barren land was tundra, grassland, and low vegetation that was highly stressed by drought. Loess and sand moved to the north and then to the south in the same way as in the northern Hemisphere. This went on for thousands of years and built up large loess and sand dune deposits.

Australia has been isolated from all other continents for over 120,000,000 years. Plant evolution in Australia took a different path from evolution on the other continents. Because of poor soils and a drier climate, life for plants was harsher than on the other continents. Drought reduces plant leaf area and causes early leaf drop, which restricts atmospheric carbon dioxide intake. This suggests that the Australian plants 28,000 years ago may have had a higher carbon dioxide requirement.

During the early glacial melting on the East Antarctica ice sheet about 26,000 years ago, Australia was surrounded by very cold oceans with some floating ice. The cold ocean water took carbon dioxide from the atmosphere surrounding Australia. This reduced the atmospheric carbon dioxide available for plants, leading to the early loss of vegetation and the large animal extinction.

In addition, there was little evaporation from the oceans to put moisture into the atmosphere. The land areas of Australia were hot from the sun's heat input. The atmosphere over Australia was also very hot, so it could carry far more moisture than was available. Consequently, very little rain fell and almost all of Australia experienced severe drought. With cold air over the oceans and hot air over the continent high winds, dust and sand storms were the result. The loess deposits in Australia didn't have the large glacial content as those in North America, Europe, Asia, and South America. However, they are believed to be present. The combination of low atmospheric carbon dioxide and severe drought greatly reduced the vegetation, which led to the large animal extinction.

As the glaciers melted, the oceans rose. By the end of the glacial breakup phase, the oceans had risen about 350 feet total. This drowning of the continental shelves flooded some of the best plant habitat available during the previous 50,000 years. Throughout the world, the lack of carbon dioxide and precipitation greatly reduced plant growth. The plant volume was at the lowest it had been since the large extinction at the end of the Cretaceous 65,000,000 years ago.

Some Plant Life Changes Associated with Extinction

Today there is great plant diversity in the tropical rain forests and much less plant diversity in the other land areas of the earth. The plant diversity on land diminishes moving away from the equator towards the poles. An example of the difference in plant diversity in the low and high latitudes is demonstrated by comparing the number of tree species in a South American rain forest to the number of species in a temperate rain forest in Alaska. In one ten acre area of tropical rain forest in South America, scientists counted 179 different tree species. Conversely, in the Tongass National Forest in southeast Alaska, 500 miles long and 100 miles wide, there are only ten native species of trees. Baranof Island in the Tongass National Forest is 100 miles long and twenty miles wide and has only two native species of trees: the Western hemlock and the Sitka spruce.

However, this situation appears to be reversed in the oceans. The most prolific life occurs in the cold oceans close to the North and South Poles. This would suggest that temperature alone was not the main cause of diversity.

The tropical rain forests have had a relatively stable environment for many millions of years. The atmospheric change during the last glacial breakup had a much greater effect on the middle and high latitudes, with milder effects near the equator. This stable environment in the tropical areas allowed evolution to work for millions of years, filling each niche in the ecosystem with new species. The location, areal extent, and composition of the tropical rain forests may have changed somewhat as the atmospheric carbon dioxide levels changed and the sea level rose and fell. There was some advancing and retreating of plant life from the mountains and uncovered continental shelves. However, most of the tropical rain forests have stayed in one place with a fairly stable environment.

Even during the glacial breakup phase, more water evaporated from the oceans near the equator than anywhere else. Much of this moisture fell on the tropical rain forests. Although the moisture was less than any other time in the glacial cycle, it was sufficient to maintain many of the plant species. The atmospheric carbon dioxide in many tropical areas was sufficient to maintain most of the rain forest plants, but not in the higher altitudes.

The end result of the rapidly changing environment, extensive ice coverage, lack of carbon dioxide, and lack of water for plant growth, was a severe reduction in plant species in the middle and high latitudes. After millions of years of instability, conditions for life have improved in the roughly 12,000 years since the end of the last Ice Age.

Plant Changes since the End of the Last Ice Age

Not long after the end of the glacial breakup phase, conditions for plant growth started to change for the better. The surface of the oceans started to warm and carbon dioxide began to come out of solution, enhancing plant growth. In addition, with the surface of the oceans warming rapidly, more water was evaporated and precipitation increased over the land. This increased precipitation was advantageous to plant growth.

The increase in atmospheric carbon dioxide and increased precipitation allowed plants to flourish and expand over the last 12,000 years. Plants migrated onto land that was previously barren. The loess and sand left on the land and the broad river flood plains, deposited as the oceans rose, were nutrient-rich, and plants grew in abundance on these lands. Trees and other vegetation migrated into many areas where glaciers had been.

The carbon dioxide and precipitation have fluctuated for the last 10,000 years, but the conditions for plant growth have been generally acceptable over this time. We do not know when humans and the animals they tended began to alter the plant community, but the effect was relatively minor up until about the last 200 years. At that time, humans learned to mechanize many functions, and the effects of human activity on the plant community rapidly became more severe. Next, we will look at the large animal extinction in detail.

CHAPTER 9. LARGE ANIMAL EXTINCTION ASSOCIATED WITH GLACIAL MELTING

Animals have certain requirements for survival, and changes in the glacial cycle altered the availability of these requirements. Each animal needs a habitat with a specific temperature range, oxygen, food, and fresh water for survival. Water and food are the most variable requirements for survival. The food supply for animals is dependent upon the availability of adequate and acceptable vegetation. When there are changes in the animals' habitats, water supply, or the vegetation available to eat, then they adjust to the changes, migrate to better conditions or die. All three of these results were common during the changes in the glacial cycle.

To put the animal changes during the glacial cycle into perspective, we'll look briefly at the origin and evolution of animals throughout geological history.

ORIGIN AND EVOLUTION OF ANIMALS

Scholars originally divided geological history based on the evidence at hand, namely groups of fossils. Precambrian seas dating from the origin of the earth to 570,000,000 years ago produced no animal fossils. Marine animal fossils appeared after that time, in the Cambrian period, and have been abundant ever since. From the Cambrian period onward, time periods were defined mainly in accordance with groups, or suites, of fossils unique to each period. Changes in suites of fossils were usually caused by extinction of animals at the end of a period. During the next period, the animals that survived the extinction evolved and produced new species to fill the gap left by the extinction. Throughout geological time, there have been five major and many lesser times of extinction affecting both marine and land animals. Over the past several hundred years, paleontologists have found sufficient fossils to establish suites for each period and to identify each extinction. However, these fossils are only a sampling of the animals that lived during these times.

Because nature has very efficient systems to recycle the materials of dead plants and animals, only a very small fraction is fossilized. Of the plant and animal remains that have been fossilized, very little are at the surface of the earth. Most have been either eroded or buried deep in the earth. Thus extremely few individual animals or plants that lived are fossilized and on the earth's surface available for study.

Some environments are more likely to produce fossils than others. Reefs and areas of rapid burial sometimes preserve many fossils in the oceans. Swamps, lakes, riverbeds, and caves sometimes preserve fossils on land. Marine animals with hard skeletons or shells and large land animals are most likely to be fossilized and preserved. On land, insects, plants, and small animals are far less likely to be fossilized and preserved. With time, fossils in the rocks can be destroyed by recrystallization. Many animal and plant species that are composed of soft tissue and no hard parts are almost never fossilized.

We are not sure when animals first appeared on earth because the first animals were very simple marine creatures that had no parts to be preserved. The first animal fossils are found in rocks about 570,000,000 years old at the start of the Cambrian period. During the Cambrian, animals rapidly became abundant in the oceans. About 410,000,000 years ago, in the early Devonian period, critters like cockroaches, scorpions, and spiders adapted to breathing air and started living on the land.

In the late Devonian period, about 375,000,000 years ago, lung fish made their way onto dry land. The amphibians evolved, living part time in water and part time on land. The emergence of animals on land occurred because there were plants to eat. In addition, the atmospheric conditions for animal life on land were becoming more favorable because there was sufficient oxygen. The makeup and thickness of the atmosphere reflected or absorbed many of the harmful rays of the sun.

In the late Pennsylvanian period, about 300,000,000 years ago, the first reptile appeared on land. Reptiles were more agile and not as closely tied to the water for survival as amphibians. Much of the earth was probably warmer than today, so cold-blooded reptiles and amphibians thrived.

Chapter 9. Large Animal Extinction Associated With Glacial Melting

In the Mesozoic era, from about 240,000,000 to 65,000,000 years ago, dinosaurs were the dominant land animals. Although dinosaurs came in all sizes, many genera were very large and consumed vast quantities of vegetation. Some of these were about 100 feet long and had massive bodies. Their fossils are often found close together, suggesting they were herd animals. Herds of large animals required fast-growing vegetation for food. High levels of atmospheric carbon dioxide during the Mesozoic would be needed to allow for this rapid growth.

Mammals first appeared during the late Triassic period, about 210,000,000 years ago. For the rest of the Mesozoic, they were relatively small, insignificant creatures that could not compete with dinosaurs. But when dinosaurs became extinct, at the end of the Cretaceous period, they left a large ecological gap. Land mammals then became larger and more numerous with many new species. Mammals have been dominant for the last 65,000,000 years.

Fossil evidence says several things about evolution: First, evolution does not occur at a uniform rate. Second, evolutionary processes respond to the existing environment and to the needs of the organisms that are evolving.

Evolution may be very slow or very fast. Cockroaches, spiders, and scorpions have remained fairly similar for the last 400,000,000 years. The coelacanth, a primitive deep Indian Ocean fish, is little changed from 70,000,000-year-old Cretaceous fossils of the Coelacanth. Coelacanths first appeared 360,000,000 years ago. Alligators are also little changed since the Cretaceous. These organisms and many others have structures that are still effective in the current environment, and so there is no reason to change.

Evolution can be very fast. After each episode of extinction, evolution progressed at a rapid rate to fill ecological niches that were left vacant.

Evolution gets feedback from the environment, which spurs what may be called specialized evolution. There are many examples of specialized evolution in nature today. The anteater's size, long narrow snout, long tongue, and internal changes are designed to let it live almost exclusively from eating ants. The marine iguana of the Galapagos Islands is the only lizard that regularly swims. It had to go into the ocean to get food. It takes a lot of evolutionary adaptation for a non-swimming desert reptile to start feeding at the bottom of the ocean. This happened on a larger scale when mammals went into the ocean. Whales and dolphins are descendants of land mammals, and they developed many complex systems so they could survive in the oceans.

There are also examples of parallel evolution. One of the most striking is the North American wolf and the Tasmanian wolf. The Tasmanian wolf looked similar to the North American wolf except that it had stripes on its hind quarters. Both were carnivorous predators occupying the same ecological niche. However, the Tasmanian wolf was a marsupial. Unfortunately, the Tasmanian wolf became extinct in 1936. This is a fine example of parallel evolution filling a necessary ecological niche. Parallel evolution has occurred repeatedly throughout the history of plants and animals.

Animal Life in the Pleistocene

Animal life during the Pleistocene era in many areas of the world was much different than it is today, because of the mass extinction of large animals near the end of the Pleistocene. In the many glacial cycles of the Pleistocene, some animal genera evolved while a few became extinct. The glacial cycles put stress on many genera.

In the last glacial cycle, the most advantageous time for most land animals was near the start of the last accumulation phase. The land was relatively warm, there was plenty of precipitation, and the vegetation was lush. Animals had plenty of food, water, and space. This was a short lived interval, however, as the living conditions for many creatures progressively deteriorated throughout the accumulation phase.

Early in the accumulation phase glaciers started to form in many places in the high latitudes and high elevations. The glaciers displaced or buried the critters previously living on this land. As the glaciers formed on the steppe and tundra environments, the grazing animals had to migrate somewhere else to survive. In North America, many of the grazing animals moved to the Great Plains. As glaciation increased, the land area available for grazing animals diminished. This reduced their population. In the same way, the glacial expansion took a great deal of boreal forest land so that the land area for browsing animals was reduced. Entire suites of plants migrated toward the equator, but the area for grazing and browsing animals was still reduced.

This same process was also occurring in Europe, Asia and South America as glaciation increased in these areas. On all of these continents, the animals were forced to migrate away from the glacial areas, which increased the number of animals living on the hospitable terrain that remained. During this time, as the volume of glaciation increased, the level of the oceans dropped, exposing the continental shelves. Animals moved to the continental shelves as soon as plants covered the new land. However, for the most part the animals that lived on the new land were different from those displaced by the glaciers because the environments were so different.

In the last half of the accumulation phase, the stress on the animals increased and survival became more difficult. With the drop in precipitation and atmo-

Chapter 9. Large Animal Extinction Associated With Glacial Melting

spheric carbon dioxide, the supply of vegetation for food and water to drink diminished. In addition, the land area covered by tundra, grasslands, and forests continued to diminish, providing less living area for animals. The situation got worse in the dormant phase.

In the first half of the dormant phase, the animals in the middle and high latitudes and the high places were living under extreme conditions. This was the glacial winter; it was very cold, with very low precipitation and low atmospheric carbon dioxide. All of these factors made life hard for most living things. More land that was barren developed in the high latitudes and high altitudes. For essentially all animals to survive, they had to migrate still farther towards the equator. The yearly winters were long and cold while the summers were short and cool. This made for a very short growing season, restricting the vegetation. Only those creatures that were well equipped to deal with the cold could survive close to the glacier. The woolly mammoth, woolly rhinoceros and other animals of the past, along with the muskox, caribou, and polar bear of today, were so equipped.

The environmental conditions that were more favorable to animals started to come back somewhat in the last half of the dormant phase. The atmospheric temperature began to rise, along with an increase in precipitation and somewhat higher levels of atmospheric carbon dioxide. These conditions brought on an increase in vegetation. The animals' supply of food and water was slowly increasing. The number of animals was slowly increasing also, and survival was not as difficult. Life continued to improve for animals in the active phase.

The start of the active phase was in the glacial spring and was a time of renewal for the animals after the long glacial winter. Plant growth became more abundant because of an increase in temperature, precipitation, and carbon dioxide in the atmosphere. This provided significantly more food and water for the animals and they multiplied. The density of animals increased over much of the land.

In the last half of the active phase, the glaciers were warmer and flowed much faster. There was abundant precipitation adding to the glaciers. With high heat input to earth there was also considerable dissipation of the glaciers, which put ice and cold water on the surface of the oceans in the higher latitudes. This cooling of large parts of the oceans' surface provided a preview of larger events to come.

The last extinction started in Australia about 28,000 years ago. The environmental conditions in the southern hemisphere leading to the mass extinction of large animals in Australia were different from the rest of the world. Dissipation on the huge continental glacier in the eastern half of Antarctica concentrated cold meltwater on the surface of the oceans around Australia. This cold meltwater absorbed carbon dioxide, lowering the atmospheric carbon dioxide over Australia. Also, this cold water in the oceans provided little moisture to the Australian continent which was very warm from the high heat input from the sun.

Drought conditions, along with lower atmospheric carbon dioxide, led to drastically reduced vegetation. These conditions were very unfavorable for survival of the large animals of Australia. Consequently, 86% of the large animals became extinct. The loss included reptiles, marsupials, and birds. These poor conditions for animal life in Australia started thousands of years before conditions became critical on the other continents. Conditions rapidly deteriorated everywhere after the start of the breakup phase.

Animal Life in the Last Glacial Breakup Phase

The last glacial breakup phase appears to have been unique in several respects that had a major impact on animal life. Earlier breakup phases were probably significantly longer, so the surfaces of the oceans did not cool as rapidly or as much. In addition, carbon dioxide levels of the atmosphere probably started at a significantly higher level, not as close to the survival limit of trees and other vegetation.

At the start of the last breakup phase there was a high density of animals on the continents living off the increased vegetation of the glacial spring. The exposed continental shelves in the low and middle latitudes were teeming with animal and plant life. Early in the breakup phase drought conditions became common, particularly in the middle and high latitudes where animals had less food and water. That was bad enough, but it was only the beginning. The tree lines and the limit of most vegetation were migrating down the mountains and out of the high latitudes because of declining atmospheric carbon dioxide. Grazing and browsing animals either migrated to other areas or died. Carnivores either migrated with their food sources or died.

In the last half of the breakup phase conditions became much worse for the animals. Many areas which previously had fostered lush vegetation became barren. High winds plagued the middle latitudes, creating large sand and dust storms. Many animals caught in these storms died from lack of food and some were buried in large loess and sand dune deposits. In addition, many animals congregated in the large river valleys draining the glacial sites and could not survive on the scarce vegetation. As the oceans rose rapidly, the river valleys became filled with water and sediment, which buried those unfortunate migrants.

Chapter 9. Large Animal Extinction Associated With Glacial Melting

The oceans were rising fast, drowning the continental shelves and the habitats of many animals that had been established over the previous 50,000 years. There may have been some genera of animals unique to the island and coastal habitats which became extinct during the breakup phase. However, in this instance the dead tell no tales because they are covered with hundreds of feet of ocean water and sediment that was eroded from the continents.

The mass extinction of large land animals occurred in North America, South America, Europe, Asia, and Australia. Smaller but significant extinction also occurred in Africa. This was the greatest extinction of large animals in a relatively short period since the dinosaurs died off 65,000,000 years ago. While the dinosaur extinction is very interesting, the extinction of the large animals near the end of the last ice age is far more important to us as humans. Most of the animals that became extinct at the end of the last ice age were large mammals. Humans are large mammals, and their population too was greatly reduced by the same conditions that caused the animal extinction. These conditions may well recur in the glacial cycles of the future.

Extent of Extinction

In the last several chapters, we saw that during the breakup phase there were barren lands between the poles and 40 degrees north and south latitude at sea level. Land down to the tree line was subject to blowing dust and sand which formed loess and sand dune deposits. This land is located between 30 degrees north and south latitude and the poles at sea level. In addition, both of these environments were found much closer to the equator at high altitudes. In many areas where loess and sand dunes were forming, large rivers in deep gorges were taking meltwater from the glaciers to the oceans. As the oceans rose, these river gorges filled with sediment.

The animal life was dying from lack of food and the effect of the dust and sand storms. When dust and sand storms became severe, animals congregated in the river gorges for some protection from the wind and to be close to the water. This is where they starved or were killed by predators. This was demonstrated in a study of mammoth remains in North America found in *Quaternary Extinctions*, 1984.. About 57% of the mammoth remains were found in flood plains of rivers and streams and in gravel pits around rivers. About 9% were found in glacial drift and outwash deposits. The rest were found in lakes, oceans, swamp deposits, loess deposits, and various other places. Most of the remains from rivers have not been found, as the animals are buried as deep as 300 feet in the old river gorges. Not only did a vast number of animals die from these severe conditions, but many genera became extinct. In addition, many genera died out on one continent but survived on another.

The very large animal extinction in North America occurred in a very short period of time, with most genera becoming extinct between 16,000 and 10,000 years ago. The diminishing atmospheric carbon dioxide and the worsening drought conditions deteriorated throughout the breakup phase and became critical near the end. About thirty-three genera of large mammals became extinct in North America in those 6,000 years while only twelve survived.

Some of these animals survived until conditions began to improve, but they were decimated and never recovered, and so they too became extinct. As the grass and vegetation returned to the land, small animals could multiply faster to take advantage of the new food source.

For example, a rabbit starts breeding at the age of six months and has a gestation period of about one month. It has two to five litters per year of four or five babies, each generating up to twenty-five offspring per year. In contrast, the large surviving grazing animals such as the pronghorn do not mate until they are over a year old and have a gestation period of about six months. A bison starts mating at the age of three years and has a gestation period of eight months. Compare this to an elephant, the mammoth's closest living relative. An elephant can bear young at fifteen years of age and has a gestation period of twenty one months. Small animals can out propagate the large animals and fill the ecological void, eating most of the vegetation before the big animals get started. In the last breakup phase, this gave the small animals a distinct advantage in surviving and expanding into the previously barren land that had new vegetation.

Some extinct animal genera have been found in what are believed to be human kill sites. Mammoth bones and artifacts have been found at sites in Arizona, New Mexico, Colorado, Wyoming, and Texas. Humans expanding into the North American heartland killed whatever large animals they found for food. Consequently, there appears to be an overlap of human migration into the interior of North America at the time of the last extinction of large animals. For several genera, humans may have been the final instrument of extinction. However, the low atmospheric carbon dioxide and accompanying drought which lasted for thousands of years were the primary causes of the large animal extinction.

In South America, forty-six large animal genera became extinct, which is more than there were in North America to begin with. This left only twelve genera surviving. In the higher elevations and in southern South America, the conditions causing the large animal extinction were exactly the same as in North America. This is highlighted by large loess deposits in southern South America. In the northern lowlands of South America, however, drought and the lower carbon dioxide levels reduced those plants that required high levels of atmospheric carbon dioxide. This drastic reduction in plant life led to the large animal extinction.

Near the end of the glacial breakup phase, conditions in Europe and Asia were similar to those in North America. Most of Russia and northern Europe were becoming barren due to low atmospheric carbon dioxide and long term drought. The evidence for this is extensive loess deposits preserved in many places throughout Europe and Asia. Fewer genera of large animals became extinct in Europe and Asia during that phase because of the vast area and the wide range the animals enjoyed. Pockets of many animals survived in the south. However, many genera of large animals did become extinct near the end of the last glacial cycle.

Africa was least affected by the reduction of atmospheric carbon dioxide and drought, although it did lose about 14% of its large animal genera. Africa was less affected because it is situated where the atmosphere is thickest, between 35 degrees south latitude and 35 degrees north latitude. Also, the drought was not as severe in central Africa because it gets much of its moisture from the intertropical warm waters of the Arabian Sea and the northern Indian Ocean, both of which had less cooling from the glacial meltwater. Still, there probably was a significant reduction in vegetation from a drop in atmospheric carbon dioxide and drought, and that would have caused the reduction in large animal genera.

As we've seen, in Australia the mass extinction of the large animals started somewhat earlier in the late stages of the active phase. Nineteen large animal genera became extinct and three survived. Life has been difficult in Australia for much of the last 65,000,000 years. The poor climate and poor soils restricted plant and large animal development.

Before the extinction, Australia had less than half as many large animal genera as any other continent except Antarctica. The large animals that were there were much smaller than those on the other continents. Low atmospheric carbon dioxide and drought caused the early extinction of the large animals. Australia was surrounded by cold meltwater from dissipating glaciers on the huge East Antarctica ice sheet. This early occurrence of low atmospheric carbon dioxide was only severe over the Australian continent.

Key Points for Extinction

We have discussed a great many things that provide a compelling argument that low atmospheric carbon dioxide was the main cause of the large animal extinction. The following is a summary of these key points.

1. Plants need carbon dioxide to survive just as animals need oxygen to survive. Land plants take carbon dioxide out of the atmosphere and expel oxygen. Land animals take oxygen out of the atmosphere and expel carbon dioxide. Throughout geological history, plants have been dominant in this relationship; the atmospheric carbon dioxide has decreased while the atmospheric oxygen has increased. The atmospheric carbon dioxide concentration is now down to about 360 parts per million and the atmospheric oxygen concentration is about 210,000 parts per million.
2. The original food source of all animals is plants. If there are no plants to eat, then animals cannot survive.
3. Carbon dioxide is thirty five times more soluble in cold water than oxygen and seventy three times more soluble in cold water than nitrogen. There is sixty times as much carbon dioxide in the oceans as there is in the atmosphere. Cold water can hold twice as much carbon dioxide in solution as warm water. Near the end of the last ice age, glacial melting put cold, fresh meltwater on the surface of most of the world's oceans. Because this water was undersaturated in carbon dioxide, it absorbed large amounts from the atmosphere. In Greenland and Antarctica, the glacial ice has been core drilled in several places. Analysis of air inclusions in the ice cores that were formed near the end of the last ice age indicate that the carbon dioxide content of the air then was about half what it is now.
4. The density of the air diminishes with altitude. Consequently, the amount of carbon dioxide capable of supporting plant life diminishes with altitude. Each plant has a minimum atmospheric carbon dioxide requirement for survival. The tree lines in the mountains are at the altitude where those trees reach their minimum atmospheric carbon dioxide requirement. The minimum atmospheric carbon dioxide requirement for most other plants is reached within about 4,000 feet above the tree line.
5. Centrifugal force from the rotation of the earth causes the atmosphere to be about twice as thick over the equator as it is over the Poles. With more atmosphere near the equator, there is more carbon dioxide near the equator to support plant and animal life.
6. Natural forces tend to concentrate higher atmospheric carbon dioxide near the equator. Warming ocean water releases carbon dioxide near the equator, and cooling water absorbs carbon dioxide closer to the Poles. Because carbon dioxide is about 65% heavier than air, over time

centrifugal force tends to move carbon dioxide in the air toward the equator.

7. There is great plant diversity near the equator where the atmospheric carbon dioxide concentrations remained higher. There is a lack of diversity among land plants in the high latitudes and high altitudes. This is because at the end of the last ice age, low atmospheric carbon dioxide killed off most of the plants. Some plants close to the equator had a high carbon dioxide requirement, and they, too, did not survive the low atmospheric carbon dioxide.

8. Massive loess and sand dune deposits were formed in the middle latitudes near the end of the last ice age. The cause of these deposits was the low atmospheric carbon dioxide and drought that accompanied the glacial melting near the end of the last ice age. These conditions killed most of the plants in the middle and high latitudes. In many areas high winds over barren and sparsely vegetated lands formed these loess and sand dune deposits.

9. The distribution of the atmosphere and atmospheric carbon dioxide is consistent with the distribution of extinction. In general, the areas that had the lowest percentage of large animal extinction were near the equator, where the atmosphere was thickest. The most severe large animal extinction occurred in the middle and high latitudes, where the atmosphere was much thinner.

10. The extinction of many large animal genera at the end of the last ice age was preceded by a reduction in size of many animals. This suggests that there was a gradual reduction in the available food supply for these animals. This is consistent with the theory that reduced atmospheric carbon dioxide caused a reduction in vegetation.

11. Most of the animal genera that became extinct at the end of the last ice age were large animals. It takes far less food to keep a viable population of small animals alive. Also, small animal populations recover much faster because they breed much faster. Many genera of small animals could survive on marginal areas closer to the equator.

12. Some suggest that the large animal extinction was caused by humans. However, physical evidence indicates that humans were victims, too. Most of the extinction in North America and northern Asia occurred before there was significant human presence in these areas, according to the evidence. Artifacts indicating the presence of humans are associated with only 20% of the large animals that became extinct in North

America. For 80% of the large animals that became extinct, there is no evidence that they had any association with humans.

Animal Life after the Extinction

After the glacial breakup, there appears to have been a lag in the low levels of atmospheric carbon dioxide and in drought conditions. In some places, the oceans were still cold and continued to absorb carbon dioxide. In other places, the oceans warmed rapidly and released carbon dioxide. By about 10,000 years ago, atmospheric carbon dioxide had increased significantly and conditions for animal life began to improve dramatically. Vegetation became lush, and fresh water became abundant. The population of surviving animals increased rapidly, and they migrated into previously barren and glaciated lands. With few exceptions, most of that animal life flourished up until about 600 years ago.

Some of the most striking exceptions were found on islands, where this flourishing was often cut short by the arrival of humans. In many cases, when humans went to islands for the first time, many of the native animals soon died out. Not only did humans kill the wild animals for food, but animals that accompanied the humans often did much damage. When people migrated, they often took along rats as unseen and unwanted guests. These rats ate birds' eggs, which decimated bird populations on many islands. Domestic animals such as pigs sometimes escaped to the wild and they, too, affected the native fauna.

For example, the moas, which were large flightless birds on New Zealand, disappeared within about 300 years after humans arrived. They were probably hunted for food by the first natives. On Madagascar, the elephant bird and many species of lemurs disappeared after humans arrived.

In the last 600 years, "the Little Ice Age," from AD 1450 to 1850, created dislocations of some animals and put environmental stress on others. Wild animal populations may have diminished somewhat because of the effect of "the Little Ice Age." A much greater impact on the animal population was caused by humans. The grazing of domestic animals took a very large area previously occupied by large herds of wild animals. Likewise cultivation, to feed the rapidly expanding human population, took away a large amount of land where wild animals had previously lived. Finally, the development of cities, towns, factories, and roads displaced or destroyed many native animals.

Chapter 10. The Human Condition during Extinction

To appreciate the human condition during a period of extinction it is important to look at human development leading up to that period. Humans and earlier human-like species have only been on earth for about 4,000,000 or 5,000,000 years. Evidence indicates that for the first 3,000,000 or 4,000,000 years of humanoid existence, they lived in Africa and had a fairly stable environment. A little over 1,000,000 years ago, some early people migrated out of Africa into Europe and Asia, and from then on humans had to cope with glacial cycles.

When the glaciers were at maximum development in the glacial cycles, life was extremely difficult for animals without fur. Those humanoids living in northern Europe and Asia had to adapt to harsh conditions. For those living close to the areas subject to glaciation, it was not necessarily the fittest but the smartest that survived. The people who developed survival strategies lived and the others did not.

The many periods of glacial stress made for rapid positive mental evolution in humans. One of the survival strategies was to migrate to milder climates in the glacial periods and expand away from the equator in the interglacial periods. These many migrations periodically brought people from other parts of the world together and also brought smarter genes to the entire population. Thus, the intelligence of the human population increased rapidly. Now we will look at the origin and early development of humans.

Origin and Development of Early Humans

What we know of the origin and development of man's earliest ancestors comes from the study of fossils; however, fossils are not representative of ancient populations. They are not found where most individuals lived. They are found where they were most likely to be preserved.

Humans are primates, and the earliest primates lived in Africa about 60,000,000 years ago. Early human-like beings were believed to originate in the rain forests of Africa about 4,000,000 or 5,000,000 years ago. Since rain forests are efficient recyclers of organic remains, the fossil evidence for this is not strong. Many anthropologists believe that humans have a common ancestor with chimpanzees and gorillas, and that they all originated in the same environment in the African rain forest.

Fossils of humanoids became more abundant in rocks less than 4,000,000 years go. This suggests that some of the humanoids started migrating away from their place of origin into other environments. They stood and walked upright, leaving their arms and hands available for other work. The use of hands for many tasks not previously undertaken started new, more sophisticated mental processes. Crude tools have been found that the early humanoids crafted about 2,000,000 years ago. The Pleistocene glacial cycles also started about 2,000,000 years ago.

Humans in the Pleistocene

At the start of the Pleistocene Epoch, all ancient human ancestors lived in Africa so the glacial cycle had only minor effects on them. Still there were effects, such as lowering and rising sea levels, that changed drainage of the land and some aspects of the climate. By 1,500,000 years ago, humans had developed the most efficient skeletal structure for functioning on two feet. This made them far more mobile and capable of migrating long distances. It also increased their capacity to function effectively in environments other than tropical rain forests.

There was probably a very rapid expansion of the human population during the interglacial periods of this time. Human remains that are about 1,000,000 years old have been found in Java. It is a long way from Africa to Java. This may suggest that there was a fairly large total human population, as the strongest motivation to move such a long distance might logically have come from a sense of crowding, in other words competition for resources. The move from Southeast Asia to Java had to occur during a glacial period when there would be a land bridge to Java.

Chapter 10. The Human Condition During Extinction

Fossil humans 800,000 years old have been found in caves in China. The climatic changes of the glacial cycles had a major effect on people living in this part of China. The alternating glacial and interglacial periods provided climate change for many people from about 800,000 years ago to present. The many times of migration and adaptation to changing environments required increased brain function, so the human brain became bigger and more complex. In the interglacial periods, life was good; the humans multiplied and migrated into the far reaches of Africa, Europe, and Asia. During each glacial period life was harsh, so only those who adapted to the new conditions or those who migrated survived. Those who adapted to colder climates needed sufficient brain capacity to make clothes, find or build shelter, and in many cases use fire.

Those who migrated had to compete with the humans already living on the migration route. The human population decreased with deteriorating climate conditions, when there was less land capable of supporting people. As people migrated back toward the tropical areas, which still had the largest concentrations of people, they also brought any evolutionary changes in their genes back to the main population.

Those people who stayed put and adapted to the new harsh conditions of the glacial period sometimes became isolated, allowing unique evolutionary changes to occur. These new characteristics may have been in size, different facial features, a somewhat different skull shape, or different brain size. When they were introduced into the main population through inter-breeding, some of these characteristics may gradually have found their way into the main population. The isolated population may have been totally assimilated by the main population and ceased to exist as a unique population.

I believe that this is what happened with Neanderthals. They started evolving as a unique isolated population about 200,000 years ago. They became shorter, stocky, and large-boned. They had heavy brow ridges, broad faces, large noses, and larger lower jaws. Neanderthals existed as an essentially isolated population until about 40,000 years ago, when the main population of modern humans migrated into their part of the world. By about 35,000 years ago, Neanderthals were assimilated into the main population and were no longer a unique population.

Some anthropologists believe the Neanderthals became extinct and were replaced by modern humans. I do not share that view. have perceived major differences in human fossils in the past, which led to many species designations. These are differences in the size of individuals, the shape of the skulls, and the size of the brain cases. However, these differences do not seem nearly as great after you have seen a dog show. Dogs have all different sizes, shapes, and brain capacity. If we found fossils of a Chihuahua, a greyhound, and a Saint Bernard, we would be inclined to believe that these were three different species. This, of course, is a special situation because humans have controlled the breeding of dogs to a certain extent. Nevertheless, dogs show the diversity that can be achieved in a single species in a relatively short period of time.

Some anthropologists use different levels of tool-making to separate species, as between Neanderthals and modern humans. However, in the twentieth century, all living humans were of the same species, but some lived in the rain forests using 10,000-year-old technology while others built rockets and went to the moon. This difference is based on different needs, motivation, and accumulated knowledge. It is not based on different mental capabilities, indicating different species.

Over the last million years or so, humans have undergone continuous evolutionary changes which led to the development of different survival strategies. For example, there is some evidence that humans used fire as much as a million years ago, but most of the identified hearths are from the last 300,000 years. By about 200,000 years ago, humans had developed to almost the same brain capacity as today.

About 60,000 years ago, humans arrived in Australia. This required a sea voyage from Indonesia of more than sixty miles. This would suggest that the people who made the voyage were proficient sailors and probably fishermen. By 30,000 years ago, humans had migrated to some of the Pacific islands east of New Guinea. This required much longer and more complex sea voyages.

About 30,000 years ago, humans crossed the Alaskan-Siberian land bridge into North America. There is little agreement on the timing of this crossing. It is probable that many of these people lived mainly off the sea. They may have been blocked from moving too far inland by the Cordilleran ice sheet, the mountains, and the mountain glaciers of the coast ranges. They may have reached as far south as South America. People who lived off the sea ate fish and sea mammals. They moved down the coast, occupying mainly what are now the continental shelves. Essentially all trace of this migration is under water since the level of the oceans has risen.

Some people started adorning themselves with personal ornaments about 30,000 years ago. Portable art such as decorated spear throwers and figurines carved from bone, wood, and ivory also became fairly common after 30,000 years ago. Art on the walls of caves has been found and dated to about 23,000 years ago. Much of the cave art that has been found is concentrated in southern France and northern Spain, although isolated instances of cave art also has been found in other areas of Europe. Artistic evidence has been found in Africa, Australia, and

Asia, dating to a slightly more recent time. Art seems to have appeared in North America and South America even later.

In the many glacial cycles of the last half of the Pleistocene, the presence and dominance of humans in Europe and Asia had increased and decreased with the glacial cycle. Because they had to cope with the stress of glacial cycles, they evolved mentally into far more effective beings. We will take a closer look at how humans fared before the glacial breakup and the large animal extinction.

Humans in the Active Phase Prior to Extinction

For those people living in Europe and Asia during the active phase of the last glacial cycle, living conditions were relatively good. During much of this time, most of the major glacial areas in the northern hemisphere were expanding to the south because of much greater precipitation.

In the active phase, there was relatively rapid expansion of humans in Asia. During this time, people reached northeastern Siberia. In addition, people are believed to have been living on part of the large land bridge between Asia and Alaska. The long ocean voyages people took to migrate to Australia and the Pacific islands early in the active phase indicates widespread seafaring skills. This suggests that many of the first humans coming to America were Asians who arrived by boat. They probably lived off the sea and had settlements on the thin continental shelves.

People probably did not move into the interior until the oceans rose and flooded out the villages on the continental shelf. The land where these coastal settlements would have been located is now under as much as 350 feet of ocean water. This is conjecture, as clearly no fossil evidence can be shown to prove directly that people lived on the continental shelves or the land bridge. Indirectly, the supposition is supported by the fact that human artifacts are found on dry land adjacent to the continental shelves. Also, continental shelves had very fertile land which may have provided food and access to the ocean for fishermen. Both of these factors would have attracted human inhabitants.

In Europe during the active phase, migration routes opened up and modern humans from the south migrated into Europe. The Neanderthals that survived the rigors of the long glacial isolation were presumably assimilated into the expanding modern population. In the latter part of the active phase, life was not as harsh, and people had the time and energy to develop art.

By the middle of the active phase, all of greater Australia was inhabited by humans. The ocean migration had reached some of the large islands east of New Guinea. In the latter stages of the active phase, much of the Australian continent was hit by lower atmospheric carbon dioxide and severe drought as the adjacent oceans became much colder. This cold meltwater came from the East Antarctica ice sheet. The low atmospheric carbon dioxide and drought took a heavy toll on the human population as much of the vegetation and large animals died off. This was just a preview of the human devastation to occur in the breakup phase of the glacial cycle.

Humans in the Last Glacial Breakup

The glacial breakup brought rapid and severe changes to most human environments. Throughout the breakup phase, humans had to contend with severe loss of vegetation and severe drought. Humans were affected by the same conditions that killed off the large animal genera, so that the reduction in human population was extreme. In many areas, no humans survived.

In most of Europe and much of Asia drought and the loss of vegetation was extensive. Humans in these areas lost their major food source as vegetation and animals that humans ate disappeared. Without food, people migrated or died. Also, there were major and fairly continual dust and sand storms which were difficult for humans to survive. These conditions led to a drastic reduction of the human population.

Oceans were rising covering the continental shelves and displacing the people that lived there. Places which today have very large continental shelves, like Indonesia, saw the land area diminish rapidly. People had to move from the lush fertile plains on the continental shelves to the rugged volcanic mountains that remained as islands. Volcanic islands could not support the large populations that would have developed on the continental shelves.

During the last half of the glacial breakup phase and early into the interglacial period, from about 16,000 to 11,000 years ago, the evidence for human habitation in many areas diminished. As the loess and sand dune deposits increased, radiocarbon dating of wood, charcoal, and bone in many areas suggest a gap in human presence for much of the late breakup phase. People disappeared throughout most of Europe and northern Asia. In southern Asia, populations diminished. There are only a few documented sites in North and South America where humans lived during this period, so the evidence is sparse.

Early in the interglacial period, about 10,000 years ago, artifact evidence of human presence on the continents became abundant. There appears to have been a population explosion and people rapidly migrated throughout the continents. The human population has been increasing rapidly ever since.

Chapter 10. The Human Condition During Extinction

Humans disappeared in many areas and diminished in other areas between 16,000 and 11,000 years ago, for the same reasons that so many large animal genera became extinct. Plants died when there was not sufficient carbon dioxide and water available for them to survive. More plant life was smothered and blown away by the dust and sand storms. This eliminated the food source for many animals, humans included. In addition, the major dust and sand storms made it very difficult for any living thing to function and survive. Everywhere the plants died out, the animals died or migrated away. In the lower latitudes, the animals were also reduced in number because their food source was reduced.

The exception to this pattern is those humans that lived at the edge of the oceans. Some people had apparently lived off the bounty of the sea for over 60,000 years. When the land plants and animals died off in the middle and high latitudes, the ocean life was still abundant. Humans living on the continental shelves, where the vegetation was gone, lived entirely off sea life — like many Eskimo and Inuit people in recent history. Many humans also lived in the tropical areas that were not affected as much by the low atmospheric carbon dioxide and drought.

The human population was greatly reduced near the end of the last ice age, probably more so than is currently recognized. The worst times only lasted about 6,000 years. Lack of human fossils and artifacts for such a short period of time is very difficult to identify. However, it is clear that humans must have been affected by the conditions that brought on the mass extinction of most of the large animal genera. Mitochondrial DNA provides some evidence.

The mitochondrial DNA story comes in two parts. First, genetic data indicates that some time in the past there was a choke point in the human population. There was a drastic and rapid reduction in the humans living on earth. The second part tries to explain when this drastic reduction occurred. In theory, genetic change occurs at a uniform rate. This theory is based on some highly controlled experiments. However, nature is variable. If you pitch your tent on a carnotite (uranium ore) deposit, you are going to get a faster rate of genetic change. If you live in a cave with high radon content, you will get a faster genetic change. If you live above 10,000 feet in altitude, you get more solar radiation and in all of these cases and many others, you get a faster genetic change.

The only documented event that could be the human choke point in the mitochondrial DNA story is the drastic reduction of plants, animals, and humans near the end of the last ice age.

Near the end of the breakup phase, the greatest concentration of humans in the western hemisphere was probably on the continental shelves around Mexico, Central America, the Gulf of Mexico, and the Caribbean Sea. As acknowledged above, essentially no direct physical evidence has been found to substantiate this view. However, there is evidence of human activity found in many places throughout North and South America as the effects of the glacial breakup subsided. This is consistent with the theory that people migrated in all directions from the Central America continental shelves as the oceans rose to flood out their settlements. In North America, the Clovis point hunting culture first appeared about 12,000 years ago. However, about 13,500 years ago, artifacts and butchered bones of a pampean gomphothere were found in Venezuela.

Most of the humans of Europe may have been concentrated in the south, close to the continental shelves, with many people around the Mediterranean Sea. This area was sheltered from the ice floes of the Atlantic Ocean. The broad continental shelves off the coast of East and Southeast Asia were also probably areas where people thrived during the early part of the breakup phase. As the people were flooded out by the rising oceans, they migrated back into the interior of Asia and Europe.

Humans have had to adapt to almost continuous climate change since they left Africa about 1,000,000 years ago. Large populations of people had slow migrations, expanding during the interglacial periods and contracting during the glacial periods. In reaction to these changes, the human brain became larger and more complex. The human capacity to think and create has increased rapidly. The development of speech allowed people to work together. The development of written language allowed people to accumulate knowledge and pass it on from one generation to the next.

Without the glacial adversity and continuous adaptation to the changing climate conditions, our mental development may not have occurred. In addition, geological changes to the earth during the glacial cycles were critical to the formation of an environment that supports large populations today.

CHAPTER 11. FIRST BIG INCREASE IN ATMOSPHERIC CARBON DIOXIDE AND POPULATION

The carbon dioxide content of the atmosphere got down to about 180 parts per million, as measured in air inclusions in glacial ice, near the end of the last ice age. This is when the surface of the oceans was at its coldest point and the oceans had absorbed the greatest amount of atmospheric carbon dioxide. After the end of the last ice age, when the surface of the oceans had warmed, the atmospheric carbon dioxide content increased to about 265 parts per million.

This increase in atmospheric carbon dioxide fed a population explosion and a rapid increase in geographic distribution of plants, animals, and humans that will be discussed later. Now we will take a closer look at what can affect the ice core data.

The ice core data available for scientists to study is believed to be pretty good. However, ice core data is least likely to be complete when there is major melting on the glacier. Near the end of the last ice age there was melting on both Greenland and Antarctica. It has been estimated that Antarctica lost 26% of its ice volume and Greenland also had major losses.

Studying ice cores is similar to studying rock cores. In rock cores, unconformities and disconformities are difficult to find if the type of rock does not change. Unconformities occur where the top of the rock bed is eroded off before more rock is deposited. This would compare to melting on the surface of the glacier before more snow and ice are deposited. Disconformities are areas where deposition stops for awhile. These compare to times when no snow and ice fall on top of a glacier for some period of time.

Where melting occurs on the surface of the ice or where no snow falls for a number of years, there is a time gap in the ice record. When the ice is later cored, there are gaps in the data. When there is faulting or flowage, there may also be gaps in the data. The data that is produced is good, but incomplete. These gaps are sometimes difficult to identify. This does not affect the long term trends that we are interested in; however, it makes exact dating shaky.

Humans in the Current Interglacial Period

As the glacial breakup was occurring, humans living on the continental shelves were flooded out by the rising oceans. They had to migrate continuously inland to get away from the encroaching oceans.

In the western hemisphere, most of the humans are believed to have lived close to the ocean around the Gulf of Mexico and the Caribbean Sea and along the northeast coast of South America. Some may have lived near the west coast of Central America. It is likely that a major portion of their food came from the ocean. Turtle Mound on Canaveral National Sea Shore is a mound of seashell debris, about sixty feet high, left by more recent Indians. However, it is probably similar to what would be found from settlements near the end of the last ice age, which are now presumably under the ocean.

Most people of Europe probably lived close to the Mediterranean continental shelves. Africa probably had a reduced population, but they fared much better than the rest of the world.

Asians were concentrated in the south near the Arabian Sea, the Bay of Bengal, and the southeast coast around the South China Sea. There were probably a fair number of humans on continental shelves around the islands now known as Indonesia and the Philippines. In Australia, most were probably located in northern Australia and New Guinea.

Essentially, no physical evidence has been found to substantiate the view that humans lived on the continental shelves. Where the humans lived is now covered with ocean and in most places, sediment. However, humans were found in many places throughout North and South America early in the interglacial period. They also were found moving away from the continental shelves in Europe and Asia during this time. This would be consistent with humans migrating in all directions from the continental shelves as the oceans rose to flood out their settlements.

For most humans the early part of the current interglacial period was a time of major adjustment. Conditions were improving with the increased atmospheric carbon dioxide and precipitation. Humans were migrating back into previously barren lands as the last of the large animal extinction was occurring.

By about 10,000 years before present, the rapid increase in atmospheric carbon dioxide and increased precipitation brought lush vegetation to much of the land, increasing animal populations. The increase in vegetation and animal life initiated a rapid expansion of human population with extensive migration to populate the continents.

Rapid Increase in Atmospheric Carbon Dioxide and Human, Plant, and Animal Population

Historically, the world's human population has been determined on a very broad scale by its ability to feed itself. Population expands to the level that can be fed. Feeding humans is a combination of natural stuff that the earth provides; knowledge of how to utilize what is available, and technology and tools to cultivate the land, harvest, and process the food. The natural stuff that the earth provides comes first. It takes much longer to develop the knowledge and technology and make it available to the expanding population.

When the atmospheric carbon dioxide increased from about 180 parts per million to about 265 parts per million, a far greater area of the earth was capable of sustaining plant growth. At the same time, there was an increase in precipitation to make the vegetation lush. Plants expanded away from the equator into the middle and higher latitudes. The suites of plants that populated these areas were quite different from the suites of plants that were on the land before glacial breakup. The plants also expanded into some areas that had previously been covered by glaciers.

The animals that had survived the extinction lived in small bands located wherever they could find vegetation. As the vegetation increased, the animals increased. Browsers migrated into forests wherever they developed. Grazers followed the grass. The animal distribution was much different than it had been before the extinction. In North America, the twelve surviving large animal genera lived in an area about twice the size of the range where forty-five genera had coexisted prior to extinction. This made for large herds, such as the bison in North America.

The individual large animals were much smaller, on average, than before the extinction. This was compensated for by having more individuals. In Australia, only three genera of large animals survived but there were many of those three. There are now kangaroos everywhere.

AGRICULTURE

Initially, most humans lived as hunters and gatherers. However, tools have been found that indicate cultivation of fields in the Middle East about 10,000 years ago. It is possible that cultivation was developed on the fertile continental shelves prior to this time. This is suggested because cultivation appeared in many places of the world at about the same time. The fine sediments of the continental shelves in many places could have easily accepted the plow.

There are several requirements for land to be used to grow crops. First, it has to have enough water, either from precipitation or irrigation. The soil has to be fine-grained so the roots of the plants can penetrate and get nutrients from it. Weathering has to occur so the nutrients are available for the plants to use. The soil has to be deep enough to accept the plow. Most land is too rocky to plow.

Of the four major sources of fertile land — river flood plains and deltas, loess deposits, glacial till, and volcanic ash — only volcanic ash is not associated with glaciation. The fluctuating sea levels of the glacial cycles created much of the fertile land in the river flood plains and deltas. When the sea level was low, the rivers cut deep valleys throughout the land. When the glaciers melted, the river valleys filled, and deltas were formed with fertile soil. This process was repeated many times in different places over the last 2,000,000 years. Glacial till left by the receding glaciers had a large component of ground rock which, with weathering, formed fertile soils.

Large loess deposits were formed by high winds over barren and sparsely vegetated lands. Loess deposits are excellent for growing grain, and they make up the breadbaskets of the world. The major grain growing areas of North America, Western Europe, Ukraine, China, and Argentina are all found on loess deposits.

Without the land changes that occurred in the glacial cycles, we could not feed the large population that we have today. Without the glacial cycles to stress us, humans would be far different than we are today.

There is a long transition from hunter gatherers migrating to populate the continents to year-round residents cultivating the land. Many hunter gatherers never did make the transition. Those that did probably had an annual route to find food with hunting and gathering stops. Next, they probably set up permanent homes at the most productive gathering stops and sent out hunting parties from there. In addition, small animals may have been trapped around the home.

The humans probably set up their homes where there were abundant fruits and vegetables. Cultivation may have come next, using wood or bone implements to plow the ground so that weeds could be removed and fruit trees and vegetables could be planted.

Domestication of animals for food and fiber came fairly early. Sheep, goats, and pigs may have been early food and fiber animals. Later, when better and bigger plows were developed, humans trained animals to pull the plow.

The domestication of grain took a long time and a fair amount of genetic engineering by ancient humans to get the grains that are known today. Modern wheat comes from three grains: spelt, einkorn, and emmer. These have been harvested in Europe and the Middle East for over 9,000 years. Wild barley has been around since the end of the last ice age. Millet has been grown in Africa for the last 8,000 years. Rice has been a major food in the Far East for at least 9,000 years. The increase in atmospheric carbon dioxide was a major factor in the development of these grains. Knowledge and technology increased dramatically when humans started building and using irrigation systems.

Irrigation and Nutrients

Increased water from irrigation enhanced plant growth and allowed plants to more effectively use the increased atmospheric carbon dioxide. This gave a greater food output to feed more humans and domesticated animals. Irrigation was a necessary first step before early civilizations prospered.

The ancient humans started with no knowledge of irrigation, so it took a long time to learn to be effective. There were many missteps with serious consequences. Of course, with all of mankind's accumulated knowledge, we are still having irrigation failures today. Irrigation projects are a very expensive proposition. They require a great deal of time, effort, and engineering. In many areas, irrigation is used as a supplement to add water to what nature supplies. In dry areas, irrigation supplies all the water to grow plants.

Irrigation water has to have a low salt content so the salts do not build up in the crop land. If salt does build up in the soil, it has to be flushed with excess water. Then fertilizer or silt must be added because nutrients are flushed away with the salt. There has to be enough water delivered at the right time in the growing season. There has to be adequate drainage so the water does not stay on the field and rot the plants. Let's look at several ancient irrigation projects.

One of the earliest forms of irrigation starts with a dam on a stream that is dry during part of the year. During the dry season, a dam is built. The pond behind the dam fills up in the rainy season. Ditches from the pond can provide irrigation water to fields below the dam. Animals can also be watered from the pond. This simple type of irrigation has been found in many places throughout the world. It has been used for over 6,000 years and is in use today.

The earliest large irrigation project started in Mesopotamia about 6,000 years ago. Mesopotamia was located in what is now Iraq and southern Iran and the projects were between the Tigris and Euphrates Rivers. Early irrigation projects were around these two rivers, which had very large spring floods that carried lots of silt. This silt provided extra nutrients to the soil. These areas are very flat, poorly drained, and had poor soil. They were subject to drought, catastrophic flooding, silting, and salt buildup on the land.

The people in this region had problems with storing water for later use and to control floods, in addition to irrigation. Silt buildup in the canals required continuous dredging. Draining the fields after irrigation was difficult and often led to salt buildup in the soil.

Later, some very large irrigation projects were put in place to overcome some of these problems. The Euphrates riverbed is higher than the Tigris. The plain of Mesopotamia, between the two rivers, is very flat. Engineers designed the irrigation system to use water from the Euphrates. The water went into canals and the crop land on the plain of Mesopotamia and eventually drained into the Tigris River. This was the most efficient project for the area.

Over the past 6,000 years many successful irrigation systems have been implemented in Mesopotamia. However, silting of the canals and salt buildup in the fields put an end to irrigation in 2000 BC, 1100 BC and AD 1200. Essentially all of these irrigation projects were government projects. When the governments were poorly managed or broke, the irrigation projects deteriorated or were abandoned. Sometimes natural forces ruined the irrigation systems and the governments fell.

About 5,000 years ago, irrigation in the Nile Valley was primitive and on a fairly small scale. The Nile Valley irrigation is unique in that its water comes from tropical highlands in Africa. The water flows almost a thousand miles through the Sahara desert to the Mediterranean Sea. The annual flood starts in June and reaches its peak in September.

Now, the Nile Valley is cut down from the surrounding land. The Nile flood plain is only about fifteen miles wide and the river is fairly straight. In the past on the flood plain, the ancient Egyptians built large flat fields for crop land next to the river. These fields were above water level when the river flow was normal and below water level during flood stage. Canals and ditches diverted flood water onto the crop land for up to two months during the growing season. This water was drained off well before harvest time. The main crops were flax, barley, and wheat. The Nile River carried silt which was deposited on the fields to provide nutrients. The Nile irrigation projects were very efficient and provided the most water on the fields with the least amount of effort. The engineering needed to design and build effective irrigation systems was the first step in learning the basics to design and build temples, pyramids, and cities.

Early irrigation systems were also created in other areas of the world such as the Indus Valley, in what is now Pakistan. Currently, the Indus Valley has the highest density of irrigation projects in the world. China built irrigation systems early on in the Hwang Ho Valley. Probably there were smaller irrigation projects throughout the world that were not associated with large ancient civilizations.

Where there were irrigation projects, plants had all four ingredients to flourish: water, increased carbon dioxide from the end of the last ice age, nutrients from the silt, and sunlight.

Growth of Civilization

Initially, four ancient civilizations were recognized. In the Middle East was Mesopotamia, located near the Tigris and Euphrates Rivers. The ancient Egyptian civilization was located in the Nile River Valley and on the Nile Delta. Another civilization was located in the Indus River Valley in what is now Pakistan. The fourth ancient civilization was located in the Hwang Ho River Valley in China.

All of these civilizations were dependent on irrigation for their existence. Up until this time most of man's time and energy was spent finding, growing, and preparing food. (His next biggest concern was shelter.) Growing food with the aid of irrigation was far more efficient than any previous method. The first four civilizations were isolated in time. They did not start or end at the same time. However, since each civilization existed for a very long time; there were times when all four were flourishing at the same time.

These four civilizations were also isolated by geography. Mesopotamia was 800 miles east of Egypt. The Indus Valley was 1500 miles east of Mesopotamia. The Hwang Ho Valley was over 3,000 miles northeast of the Indus Valley with the Himalayas in between. There may have been some communication and minor interaction between the civilizations of Egypt and Mesopotamia. However, there was little or no communication with the Indus or Hwang Ho Valley.

Each of these civilizations interacted with the people around them and they probably had small armies to keep the "barbarians" away. However, there were no serious threats to their existence since there were no other civilizations nearby. Also, these four civilizations were living in the most prosperous areas, which met all of their needs. Consequently, they had little interest in major conquest. They probably were very inward-looking, and were more interested in building cities, temples, and monuments.

The four civilizations were located at latitudes that were marginal, at best, before the end of the last ice age. The productive Nile Valley was at about 30 degrees north latitude. The irrigated Tigris and Euphrates was at about 34 degrees north latitude. The Indus Valley was at about 26 degrees north latitude and the Hwang Ho Valley was at about 36 degrees north latitude. These areas were all very productive with the increase in carbon dioxide after the end of the ice age.

With the success in irrigation came changes in population dynamics. These populations were living in relatively small areas in the river valleys. This made the population density much greater in the areas covered by the first civilizations. Outside the regions under irrigation, the human population was spread out across the land.

More civilizations sprang up long after the first four were developed. These later civilizations were more complex and did not develop around irrigation. Several civilizations developed around the Mediterranean Sea. Increased productivity of the world's farms made the advances in civilization possible, in Asia, Africa, and the Western Hemisphere as well as around the Mediterranean.

Warm Spell before the Little Ice Age

There was a warm spell from about AD 800 to near the start of the Little Ice Age. It was called the Medieval Warm Period. The Medieval maximum of solar activity occurred during this time. On average, the world climate a thousand years ago was warmer than it is today. This is when Greenland got its name, as the coast of Greenland was warmer, greener, and more productive than it is today. The surface of the oceans were warmer than today. The edge of the sea ice was much closer to the North and South Poles.

The lack of sea ice allowed Viking ships to colonize Greenland. These colonies prospered for several hundred years. The Vikings were also able to establish a North American colony for a short time.

The warming of the surface of the oceans during this spell put lots of energy into the oceans. The oceans also gave off some carbon dioxide so that there was a small increase in atmospheric carbon dioxide.

Little Ice Age

The Little Ice Age lasted from about AD 1450 to 1850. At the start of the Little Ice Age, the surface of the oceans was warm and there was a great deal of evaporation. In the higher elevations and higher latitudes, the atmosphere was becoming cold and wet with increased snow and ice. There was much greater cloud cover from the increased water vapor in the atmosphere. More of the sun's energy was reflected off the cloud tops and snow, back out into space, making it colder. Initially, the atmospheric carbon dioxide level was a little higher than it had been up to that point in the interglacial period.

Conditions began to change. The surface of the oceans started to cool. However, the atmosphere was also cooling. This still made for abundant rain and more snow and ice in the higher elevations and higher latitudes. With an increase of snow on the ground and abundant cloud cover, a great deal more energy was reflected back into space. As the surface of the oceans cooled, some of the atmospheric carbon dioxide was absorbed.

Glaciers began to form and advance in many areas around the world. Essentially, all glaciers reached their largest area since the end of the last ice age 12,000 years ago. The best way to look at glaciation in the Little Ice Age is to start with glaciers present today and work backwards. Most of today's glaciers are left over from the last ice age or are remnants of the Little Ice Age glaciers. However, most of them have been reduced in size by 150 to 200 years of melting. We will also look at some areas where Little Ice Age glaciers have melted completely away.

Today, the glaciers in Alaska cover about 29,000 square miles. There may be more than 100,000 glaciers, although only about 650 have been named. Glaciers are complex and operate on glacier time, not human time, so it is sometimes difficult to understand what is going on.

We have the most information about tidewater glaciers. Explorers first mapped the ice front in Glacier Bay in about AD 1775. The ice fronts in Glacier Bay have been mapped many times since, recording when the glaciers receded and advanced. Essentially, all of the glaciers receded from AD 1775 to 1925. Most of the glaciers have been stable or advancing since then. The tidewater glaciers around Prince William Sound and the majority of these are stable or advancing.

One major exception in Prince William Sound is Columbia Glacier. It is a very large tidewater glacier covering 435 square miles. It is located twenty-seven miles west of Valdez, Alaska. Columbia Glacier is about seven miles east of the epicenter of the 1964 Alaska earthquake, which was 9.2 on the Richter scale. Columbia Glacier first advanced in the early 1970s and then went into extensive retreat in the 1980s. The recent Columbia Glacier action may be related to the 1964 Alaskan earthquake. Mears Glacier, located eight miles north of the earthquake epicenter, comes from the same ice field as Columbia Glacier. However, Mears Glacier has been advancing for at least 100 years.

A very large percentage of the mountain glaciers that end below 5,000 feet in elevation have been thinning, retreating, or stable over the last hundred years. Mapping methods on glaciers have changed dramatically over the last hundred years. Initially, mapping was done by people on the surface of the glacier. In the

1950s, many of the glaciers were mapped using aerial photography, so the mapping was on top of the snow. By the 1990s, glaciers were mapped using side-looking radar, which looked through the snow and mapped on top of the ice. When comparing the results of these different methods there is a certain amount of error.

Near the end of the Little Ice Age the glaciers were considerably larger than they are now; how much bigger is variable and unknown. Consequently, for other parts of the world we will focus on current ice in the glaciers.

In the western United States, there are about 200 square miles of glaciers. However, there are many features such as cirques which held glaciers during the Little Ice Age, but the ice is gone now. From these we know that glaciation was more extensive 300 years ago.

The glaciers today in South America cover about one-third the area covered by Alaskan glaciers. This is about 10,000 square miles. Most of these glaciers are in the mountains of southern Chile, between 45 and 55 degrees south latitude. New Zealand has about 300 square miles of glaciers located on the highest mountains of the South Island.

The Canadian mainland has about 19,000 square miles of glaciers. Most of these glaciers are located in the Coast Mountain Range of British Columbia. Some are located in the Saint Elias Mountains in the Canadian Yukon Territory. These are mainly in the Kluane National Park Reserve. There are also glaciers in western Alberta. In the Canadian Arctic Islands there are about 58,000 square miles of glaciers. Most of these glaciers are on Ellesmere, Axel Heiberg, Devon, and Baffin Islands. These islands are mainly west of Greenland across Baffin Bay.

Iceland has about 4,500 square miles of glaciers. This is about 11% of the total land area. Svalbard, a group of islands halfway between Scandinavia and the North Pole, has about 15,500 square miles of glaciers. The Russian Arctic Islands have about 22,000 square miles of glaciers.

Southern Europe and the Alps is the area where the study of glaciers started and it has about 1,900 square miles of glaciers. Scandinavia has about 1,500 square miles of glaciers.

The mainland of Asia has about 45,000 square miles of glaciers. Much of this is in the High Himalaya Mountains of China, Tibet, and Nepal.

In essentially all areas discussed, the land covered by glaciation today is less than it was at the height of the Little Ice Age. The rate of melting is highly variable. There has also been new snow and ice deposited in the last 150 years. This is also variable for each location. There were increases in ice volume on Antarctica and Greenland. The increases in ice volume were probably significant; however, there is no way to identify it.

During the Little Ice Age, the surface of the oceans grew colder. In the latter part of the Little Ice Age, sea ice grew and extended much farther away from the Poles than any time since the last ice age. The atmospheric carbon dioxide is believed to have diminished some because of the cold ocean surface absorbing carbon dioxide.

The Little Ice Age is believed to have been caused by a reduction in sunspots and associated solar flares. They expel vast amounts of energy, some of which reach the earth and warm it up. The sunspot activity started to diminish and the atmosphere began to cool about AD 1300. During the Maunder Minimum, between AD 1645 and 1715, there is no record of sunspot activity. After the Maunder Minimum was over, sunspots slowly returned to the surface of the sun, warming the earth. By AD 1850, the warming trend was well-entrenched, ending the Little Ice Age.

Volcanic activity may have been a contributing factor to the Little Ice Age, although these are a short term phenomenon. The eruption of Tambora Volcano in Indonesia in 1815 put large amounts of ash into the atmosphere. This ash reflects sunlight back out into space, blocking some of the incoming warmth for a short time. The year 1816 was called "the year without a summer," due to such cooling.

HUMAN RESPONSE TO THE LITTLE ICE AGE

The Little Ice Age brought cold wet winters and was very difficult for most humans. The greatest effect was on humans living in the mid to high latitudes and the higher elevations. Most of Europe is located between 45 and 65 degrees north latitude, which was an area of great stress. The weather got colder and wetter, and crop failures and famines became more frequent.

Many farms in the higher latitudes of Scandinavia were abandoned. The growing season in England was one or two months shorter than today. Different crops had to be planted to allow for the shorter growing season. In Germany, the wine production during the Little Ice Age was less than half that in the previous warm period. Advancing glaciers in the Swiss Alps overran farms and villages. The rivers and canals of the Netherlands often froze. In England, the Thames River also froze.

The general health of the human population declined during the Little Ice Age. Colder and wetter weather and malnutrition from low crop yields made humans far more susceptible to diseases from flu to malaria. This, along with famine, caused a very high death rate. The Greenland colony did not survive and the population of Iceland was cut in half.

When populations are under stress, there is usually political upheaval. In Europe, there were many wars and revolutions between 1450 and 1850, as governments tried to solve or divert attention away from the lack of food.

Successful strategies were colonization, industrialization, trade, and emigration. Colonies were established in lower latitude areas that were less affected by the climate change. The colonies took people away from the homeland, which helped offset the food shortage. Also, many of the colonies produced surplus food. The first trade was with the colonies and, in time, trade was with whoever had goods to trade. Northern European trading companies traded industrial goods for food and raw materials. Emigration was a successful strategy for many individuals. Moving to a far away land to improve one's life is a difficult decision. However, if the oldest son has inherited the family land and you are left with nothing, and have no prospects, the decision becomes easier.

Early in the Little Ice Age, Great Britain established major colonies in North America, the Caribbean, and Asia. The North American colonies in the area of the United States and Canada were mainly located between 30 and 50 degrees north latitude. There was an abundance of land capable of producing surplus food and providing a good living for many immigrants from the United Kingdom. In Canada, the Hudson Bay Company acquiring furs came long before any British government colony.

The British colonies in the Caribbean included: Jamaica, Grenada, Barbados, Bahamas, Barbuda, and Antigua. The British immigrants to these islands grew tobacco, sugar, and spices, and made rum. They provided trade goods and food to the English economy.

Emigration was a strategy used by many individuals and families in countries that were seriously affected by the Little Ice Age. Immigrants went to all countries where colonies were established, but the United States received more immigrants. From 1620 until the Little Ice Age was over in 1850, more than 2,500,000 immigrants came to the United States. At the time of the American Revolution, three-fourths of the population was of United Kingdom or Irish descent. Most of the others had come from Germany, the Netherlands, France, and Switzerland. Most of these people came to the United States because the Little Ice Age conditions drastically reduced the food supply in their home countries.

In the later stages of the Little Ice Age, there was a flood of immigrants from northern Europe. All these new people had to be fed, which provided a major motivation for inhabiting the continent. Initially there were explorers sent out by governments to map and evaluate the potential of the land. The next people moving into the unknown areas were the fur trappers and traders. Since they were dealing with the native population, their survival was based on peaceful co-existence.

We normally think of populating the United States as an east-to-west migration, but early in our history, the migration was from all directions. Spanish explorers, Mexican traders, and ranchers came from the south into Texas, New Mexico, Arizona, and California. Ships from several countries established ports on the West Coast and the migration was to the east toward the interior of the country. Trappers and traders of the Hudson Bay Company and other fur companies came from the north, giving some north-to-south migration. Much of the later migration was more east to west, but there were still migrations in all directions.

Although the effects of the Little Ice Age were receding all over the world, immigration increased rather than slowing down. There were still many people leaving northern Europe. The world was on the brink of unprecedented growth.

Chapter 12. Current Big Increase in Atmospheric Carbon Dioxide and Population

From the end of the Little Ice Age in 1850 until now, there have been massive changes to the world. Atmospheric carbon dioxide has increased by almost 30%. Humans have gone from mainly living on farms and in villages to mainly living in large crowded cites. The human population has increased by over sixfold, from about 1,000,000,000 to 6,500,000,000 people. We will explore the causes of some of these extreme changes.

Large Increase in Atmospheric Carbon Dioxide and Population in the Last 150 Years

Every time there has been a major change in atmospheric carbon dioxide during the last 20,000 years, there have been major changes to plants, animals, and humans. Near the end of the last ice age, there was a drastic drop in atmospheric carbon dioxide. This caused an extensive loss of vegetation in many areas; an extinction of over half of the large animal genera of the world; and the elimination of most people. After the end of the last ice age, there was a large increase in atmospheric carbon dioxide. This brought extensive vegetation to the previously barren lands and a vegetation pattern in the mid to high latitudes that was different from anything before; a vast increase in animal life; and a population explosion with extensive migration to many parts of the world.

There has been an extensive increase in atmospheric carbon dioxide over the last 150 years, from 280 to 360 PPM, which has caused an extreme increase in plant and animal productivity. This in turn has caused the very large increase in the human population.

Figure 8. Carbon Dioxide and Human Population Increases 1800-2000.

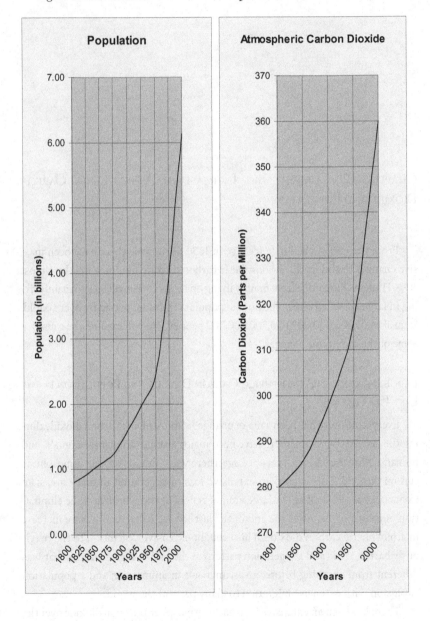

Chapter 12. Current Big Increase In Atmospheric Carbon Dioxide And Population

The initial increase in atmospheric carbon dioxide was caused by warming of the surface of the oceans after the end of the Little Ice Age. With this warming, some carbon dioxide was coming out of solution. In addition, with a warming ocean surface carbon dioxide was being absorbed at a slower rate. Much of the extensive sea ice melted, with the limit of sea ice receding back toward the Poles. No one knows how much of the increase in atmospheric carbon dioxide is natural. Most of the increase in carbon dioxide is a product from the burning of fossil fuels.

Oil, a basic fossil fuel, has been collected from oil seeps for thousands of years. However, oil was not used extensively until the Drake Well was started in Pennsylvania in 1859. There is some evidence that coal has been used as a fuel for thousands of years. In the United States small amounts of anthracite coal were used in the early 1800s. However, usage was minimal until bituminous coal mining started in about 1870. Natural gas usage started in 1872 but was negligible until the early 1900s.

On a worldwide basis coal usage started to increase in about 1850. There was a steady increase until World War I. From World War I until the early 1950s coal usage remained fairly flat. Coal was used for power plants, to fuel steam locomotives, and home heating. From the early 1950s until now, the amount of coal used has nearly tripled. Most of the coal is now used for power generation. There are no more coal-fired steam locomotives. Coal for home heating has been replaced by oil or natural gas.

Oil usage was relatively minor until World War I. Since then the proliferation of automobiles, trucks, locomotives, and airplanes has dramatically increased oil usage. Petroleum is also used for power generation and as a raw material to make petrochemicals. Oil use increased slowly from the 1970s until 2000. Oil demand from China and India has caused world oil usage to rise rapidly since 2000.

Worldwide natural gas usage started to expand rapidly in the 1940s and has expanded steadily until now. In the United States, one of the big drivers of natural gas usage was the development of the Hugoton Gas Field in western Kansas, Oklahoma, and Texas. It had over seventy trillion cubic feet of recoverable gas at a shallow depth and provided inexpensive home heating to many people. Natural gas is also used for power generation and as a raw material for many chemical processes.

As the graphs in Figure 8 show, the increase in world population has mirrored the increase in atmospheric carbon dioxide. In many studies, it is not uncommon for people to explore the obscure and ignore the obvious. Population mirroring the carbon dioxide increase is the obvious.

Most of the population growth has occurred in developing nations. Much of the knowledge, technology, and tools needed to take advantage of the excess carbon dioxide for increased food production were in the developed nations.

There are a number of ways in which the added carbon dioxide can expand the total population of the world. In the developed world, agriculture scientists improved seeds and developed many more varieties of food plants. They found more efficient ways to farm and raise livestock. Better ways to store, transport, and package food were devised to keep from spoiling or wasting food.

There was hunger in many nations in the wake of the two World Wars and the Great Depression. Consequently, most developed nations made self sufficiency in food production a top priority.

Crop Yield Increase Caused by Increase in Atmospheric Carbon Dioxide

Feeding the world's over 6,500,000,000 people today requires cultivation of fertile land. Only about 13% of the land on earth is used to raise crops and about 22% is used for grazing domestic animals. The rest of the land area is forest, desert, mountain, tundra, barren, already covered with human structures, or covered with ice.

The land that is currently under cultivation is about all that can be farmed. Most of the increase in food in the past sixty years has come from increased yields from that area. Future increases in food will also mainly come from increased yields. The increase in atmospheric carbon dioxide is the major factor in increased yields. Carbon dioxide gas is sometimes pumped into greenhouses to increase plant growth.

Over the last thirty years in China and India, rice production has doubled and wheat production is about four times larger. Better seed and fertilizer are factors related to these increases but increased carbon dioxide is the major factor.

In the United States, corn yields were 30 bushels to the acre in 1945 and 140 bushels to the acre in 2002. Some of this increase was due to more corn being irrigated.

One of the best ways to look at increased yield attributed to increased carbon dioxide is to look at long term data on the Kansas wheat crop. It is affected by fewer variables. The wheat in Kansas is winter wheat. It is planted in the fall and starts to grow before winter. In winter, the wheat goes dormant. In spring, it starts growing again and is harvested mainly in June and July. Wheat is basically a dry land crop. In Kansas, irrigating wheat is not a profitable venture. Kansas land is fertile and the fertilizer used just offsets what has been lost in the last hundred years.

Data is gathered periodically throughout the growing season on the wheat crop. The wheat crop sketches are a compilation of this data and take into consideration seeded acres; harvested acres; excerpts of wheat conditions to show what can happen; final yield per acre; total bushels of wheat production; and leading varieties of wheat.

We look at seeded acres versus harvested acres because there are many things that can kill the wheat before harvest, such as drought, hail, floods, bugs, and weeds. We show leading varieties because new varieties are needed to take advantage of the added carbon dioxide. New varieties may be more drought or disease resistant. Most varieties are replaced within a ten year period.

WHEAT CROP SKETCHES: 1918–1998
(Source: Kansas Agriculture Statistics Service)

1918 Crop: Seeded acres — 10,199,000; harvested acres — 7,250,000.

Wet snows of February and light rains in March benefited wheat, but winter kill was severe in northern and western Kansas. Final yield per acre — 13.5 bushels. Total wheat production — 97,710,000 bushels.

1928 Crop: Seeded acres — 12,761,000; harvested acres — 10,639,000.

Damage from Hessian fly in western counties became apparent late in the fall. Some loss of wheat from hail and heavy rain occurred in June, but additional rain increased yields to more than offset losses. Final yield per acre — 16.3 bushels. Total wheat production — 173,185,000 bushels.

1938 Crop: Seeded acres — 16,942,000; harvested acres — 14,494,000.

Much wheat entered the dormant period in poor condition, and below normal precipitation with high winds in January and February caused considerable abandonment. Above normal moisture in March and April, rains were favorable. Widespread infestation of orange leaf rust, and black stem rust all contributed to holding down yields. Final yield per acre — 10.5 bushels. Total wheat production — 152,163,000 bushels.

1948 Crop: Seeded acres — 14,634,000; harvested acres — 13,221,000.

Dry topsoil during the fall was unfavorable for seeding in the western two-thirds of the state. Moderate temperatures and abundant rainfall during June resulted in yields much greater than expected. Final yield per acre — 17.5 bushels. Total wheat production — 231,368,000 bushels. Leading varieties were Pawnee, Comanche, Tenmarq, Wichita, and Early Blackhull.

1958 Crop: Seeded acres — 10,727,000; harvested acres — 10,433,000.

An unusually large proportion of seedings on summer fallowed land (land that rested for a year), abundant moisture, thick stands, and cool damp filling weather all contributed to an excellent yield per acre, uniformly high across the state. Final yield per acre — 28.5 bushels. Total wheat production — 297,340,000 bushels. Leading varieties were Wichita, Kiowa, Pawnee, Triumph, Ponca, and Comanche.

1968 Crop: Seeded acres — 11,963,000; harvested acres — 9,751,000.

In extreme western Kansas, lack of moisture into early spring, coupled with greenbug and cutworm damage, caused sharp acreage losses and reduced yields. Final yield per acre — 26.0 bushels. Total wheat production — 253,526,000 bushels.

1978 Crop: Seeded acres — 11,300,000; harvested acres — 10,000,000.

Because of grasshopper damage, some field borders had to be reseeded. On April 30 and May 1, most of the western two-thirds of the state received more than one inch of rainfall, relieving drought stress. Many local areas received damage from hail and heavy rains at harvest time. Final yield per acre — 30 bushels. Total wheat production — 300,000,000 bushels. Leading varieties were Eagle, Scout, Sage, Centurk, Triumph, and Tam.

1988 Crop: Seeded acres — 10,200,000; harvested acres — 9,500,000.

Leaf rust was present over the entire state and wheat streak mosaic developed in epidemic proportions in most areas. June was hot and dry, causing rapid maturing. The Russian wheat aphid spread across the western half of the state causing additional yield losses. Final yield was 34 bushels per acre. Total wheat production — 323,000,000 bushels. Leading varieties were Arkan, Newton, Larned, AgriPro Hawk, Pioneer, and Agripro Victory.

1998 Crop: Seeded acres — 10,700,000; harvested acres — 10,100,000.

During late May, a severe hail and wind storm moved across the west central, southwest central and south central parts of the state, destroying some acreage and causing significant damage to the crop. Final yield was 49 bushels per acre.

Total wheat production was 494,900,000 bushels. Leading varieties were Jagger, 2137, Tam 107, Karl/Karl 92, 2163, and Ike.

These details suggest a few conclusions. In the 1918 crop, 28.9% of the seeded acreage was abandoned. At that time, harvest was a long tedious process. The farmer had to decide if the cost of harvesting was greater than the value of the grain that could be recovered. If the farmers had harvested more marginal, and at that time non-commercial, fields, the yield per acre would have gone down from the 13.5 bushels per acre even though total production would have gone up. In the 1998 crop, abandonment was only 5.6%.

The 1938 crop occurred near the end of the Dust Bowl period. However, they then had combines to harvest the grain and the cost to harvest was less. They could still make money harvesting lower yield crops. This is one reason the 1938 crop yield was so low, at 10.5 bushels per acre. The other reason is that far more land was seeded and harvested in 1938 than any other year. Great stretches of marginal land were seeded.

From 1918 to 1998 the wheat yield per acre increased about fourfold. Essentially all of the increase in wheat yield was the result of an increase in atmospheric carbon dioxide from about 308 to 360 PPM during those 80 years.

Other Factors Associated with Food Supply

There are many factors, both positive and negative, other than increased atmospheric carbon dioxide, that affect the food supply. We will look at the effect of some of these factors over the last 150 years as the atmospheric carbon dioxide rose. Positive effects have been produced through use of irrigation, mechanization, fertilizers, pesticides, and herbicides. But some positive factors are a double-edged sword, as the benefit may be hard to maintain over time. Many of the negative factors relate to taking land out of production.

Irrigation can and has greatly increased the amount of food produced on an acre of land. The water for irrigation comes from two sources: surface water and groundwater. In the early 1900s, irrigation came into vogue in the dry areas of the western United States. Irrigation became prevalent with the building of many large dam projects. Irrigation was further expanded in the 1940s and 1950s with the increased pumping of underground water.

Under the western portions of Nebraska, Kansas, the Oklahoma panhandle, and the Texas panhandle, there is an aquifer called the Ogallala Formation. The Ogallala is the water source for irrigation throughout much of this area. The water is pumped from this underground source and is used on fertile ground, much of it loess, to provide abundant harvests. Much of the Ogallala area in Texas, Oklahoma, and southwest Kansas is underlain by the Hugoton gas field. This was the largest gas field in North America and in the past, it provided cheap fuel to run the irrigation pumps. For about forty years, there was cheap fuel, abundant water, and fertile land; this area produced about 16% of the food raised in the United States.

Now there are several flies in the ointment. The cheap natural gas is a thing of the past. There is very little recharge of the Ogallala aquifer going on. More water is being pumped out than is seeping back in, by far. The water level is being pumped down, and some wells are drying up. The loss of irrigation in the Ogallala represents a significant drop in the food-productive capacity of the United States.

With water depletion, irrigation is not an operation that can go on forever. Several other things can go wrong. As cities expand, they need more water and often this water is diverted from irrigation projects. Also, sometimes surface water supplies become too salty for irrigation.

When I was in the oil business in the 1970s, I had to travel through a small town in West Texas that was almost a ghost town. The countryside was desolate, with abandoned irrigation canals around the country. One day I stopped to eat my lunch at a lake near this town and I noticed there were the white coatings of salt on the shore of the lake. I talked to a native of the area, who confirmed that the lake, with time and high evaporation, had become salty and could no longer be used for irrigation. Without the water, and the original vegetation being gone, the area was very barren and slow to recover to its natural state.

These same processes are going on throughout the world. Agriculture is using up some water resources and other resources are becoming salty. There is greater competition for water, and clean pure water is becoming a highly prized commodity. In some farm areas with small numbers of people, more of the water is appropriated or purchased by the cities. There are few identified opportunities for new irrigation projects. In the future, the world food-productive capacity from irrigated lands will go down as more water is diverted to cities. This will put much greater pressure on the land because the food will have to be produced without as much irrigation.

In many areas, the use of farm machinery has increased dramatically over the last 150 years while there has been an increase in atmospheric carbon dioxide. The countries that use modern farm machinery extensively are the United States, Canada, United Kingdom, Australia, and the countries of Western Europe. Many more countries are made up of a mosaic of small plots of land, farmed in the old traditions, mixed in with large farms using modern machinery. Over half of the world's land is still tilled under conditions that do not permit the use of modern farm machinery. This includes most of the small plots of land in Southeast Asia that grow rice which has to be flooded.

One of the big advantages of mechanization is that it saves manpower. This is a blessing in countries that have more jobs than people. Farm machinery does the job much faster than people and animals. A hundred years ago, the wheat harvest lasted about two and a half months. The wheat was in the field subject to weather, which cut down the yield. Today with a modern combine, harvest is usually complete in less than two weeks and there is little deterioration from weather. Production from marginal land is also far more economic using farm machinery.

During the last half of the 1800s, the farming areas of the United States were populated by immigrants who had been farmers in Europe. The European land had been farmed for a long time and had lost many of the nutrients in the soil. Manure was about the only fertilizer used and the farm population was greater than the land could sustain, so many came to the United States.

This same pattern occurred in the United States when the nutrients were depleted in the thinner eastern soils; many people just sold out their farms and moved farther west. In the early 1900s, the migration slowed down because there was not much good well-watered land left to break out with the plow.

Farmers had to start putting nutrients back in the soil. They first did this by rotating their crops. Next, they started using fertilizer to replace the nutrients. Feeding 6,500,000,000 people in the world requires intensive farming, and that means using greater volumes of fertilizer. The increased plant growth made possible by increased carbon dioxide and water requires an increase in other nutrients as well.

For about sixty years, pesticides have been used in significant amounts. They have improved the quality of many crops and of livestock. Spraying or dusting a crop against an infestation of bugs often increases the volume of food harvested. These good effects often make the use of pesticides economical. There is also a downside to pesticides. They kill beneficial insects along with the bad insects. Also, the birds and small animals that eat insects often have nothing to eat, and they die. We don't know all the effects of the loss of beneficial insects, but one problem is that a decrease in the population of honeybees and other beneficial insects prevents proper pollination of the crops. Some farmers will not use pesticides for this reason.

Herbicides are often used to kill weeds in a crop. Usually, this is done when the plants are young and susceptible. Usually, the crop is harvested before the weeds get high. If the harvest is delayed by rain, then the weeds can take over and ruin the crop.

Herbicides are used with low-till farming methods. With most farming, the ground is plowed to kill the weeds before the crop is planted. In low-till farming the vegetation on the ground is killed with a herbicide and then the new crop is planted. This eliminates one plowing and the land is more resistant to wind and water erosion.

Fertile soil is lost on most cultivated land throughout the world because of water and wind erosion. On a one-year basis, this loss is often hardly noticed and is of little consequence. However, the loss is cumulative, and over many years, the destruction of the land is severe. This problem is of serious concern today, particularly in Africa and Asia.

Soil erosion from wind and water occurs if people have messed with the land, and it occurs if people have not messed with the land. It just happens a lot faster if people are involved. The major ways the land is damaged is by the plow and farming in general, plus deforestation and overgrazing. Humans destroy all or parts of the plants' root systems that hold the soil in place, so wind and water can work on the soil. In the last sixty years, there has been serious damage to that 13% of the earth's surface that is covered by crops. There was damage before then, but no one was keeping track of it. In the future, the problem of soil erosion will continue to expand, destroying much more crop land. This will make it far more difficult to feed the expanding population.

Currently about 13% of the land on earth is under cultivation and 22% is pasture. The land under cultivation is about all the land that can accept the plow and has enough water to grow crops. The land under cultivation is also an attractive place to build cities. The rapid population growth that has occurred since the end of the Little Ice Age means that many new people are living in cities built on land that was previously cultivated. Future expansion of these cities will take land out of cultivation. Furthermore, some land used to grow food is being diverted to grow plants for ethanol production. When crop land is diverted from food production to other purposes, there will be less food to feed the population and some people may get hungry.

Factors Altering Human Population Growth

The human population has grown from about 1,000,000,000 to over 6,500,000,000 people in the last 150 years. The 28% rise in atmospheric carbon dioxide, which increased the food supply, accounts for most of the rise in population. However, other factors have increased or decreased human population

Chapter 12. Current Big Increase In Atmospheric Carbon Dioxide And Population

growth. Most of the recent population increases occurred within 30 degrees latitude north or south of the equator. Most of these population increases came from high birth rates.

In areas near the equator, it was not as difficult for the young to survive. Goods did not have to be stored up or solid houses built to survive the winter. Wealth was far less important. Life was much simpler, and there was a different mindset. There was no need for an extensive education. Consequently, a much higher proportion of the population started families at very young ages. Without education, they had much larger families.

In the mid and higher latitudes, the population increases were much less, and a significant amount of the recent increase was caused by a longer lifespan. In the developed countries, the birth rate is much lower than the emerging countries. As fewer babies are born and the living population survives to an older age, on average the developed countries have a larger population of older people.

Improved health care is fairly costly and only the wealthy nations have been able to pay the high cost. Still, many more people around the world are being immunized against the common diseases. Greater health care worldwide relates to a longer life span, which will further increase the world population.

Wars decreases the world population. In the past, most of the people killed in wars were the people that were fighting. The two sides engaged in the battle, and that is where the killing took place. However, in World War II, large numbers of civilians were killed. A total of about 51,000,000 people were killed, including 15,000,000 combatants and 36,000,000 civilians. When people's lives are disrupted and there is uncertainty, then birth rates fall. There have been many regional wars which have together altered human population growth.

There also have been a number of depressions worldwide during the last 150 years that lowered the birth rate.

Natural catastrophes such as hurricanes and typhoons, tsunamis, earthquakes, volcanoes, and lightning have killed millions of people. As the world population has increased, more people have chosen to live in high risk areas. The coastlines are very popular despite the risks of storms or tsunamis. Little thought is given to living in areas susceptible to earthquakes or volcanoes.

Epidemics of infectious diseases such as flu, tuberculosis, malaria, smallpox, and AIDS have affected world populations. All forms of birth control have affected birth rates in many countries. This is necessary if there is going to be control of world population growth.

The Oceans Rising or Seashores Sinking

Often we hear the oceans are heating up and expanding or that glaciers are melting and the oceans are rising. This is determined from measurements of minor sea level changes on the oceans' coasts and from flooding in places like Venice, Italy. However, a pretty good case can be made for just the opposite, that the oceans are not rising but the shorelines are sinking.

From a geological point of view, let us look at what happened on the coastlines of the world before humans had an impact. The pebbles, sand, silt, and clay found at the coastline did not originate there, but came from the interior of the continent. As an example, we will look at the origin of the sediment on the northern shore of the United States Gulf Coast. Most of this sediment came from the rivers that flow into the Gulf of Mexico from the Rio Grande in south Texas to the Alabama River flowing through Mobile Bay. This sediment came from the mid-continent area between the Rocky Mountains and the Appalachian Mountains. To narrow the example further, we will look at what came down the Arkansas River to the Mississippi River and on to the coast.

The Arkansas River starts high in the Rocky Mountains near the Continental Divide at over 10,000 feet in elevation. As it goes through the Rockies, there are many rapids where large boulders are moved and broken up. Water coming down steep canyons is a powerful force to break up rocks and move sediment. As the Arkansas River got to the plains south of Pikes Peak, it was at an elevation of about 5,000 feet and was carrying lots of cobbles, pebbles, sand, and silt. In crossing the plain, it ground its load of sediment into smaller particles. Streams flowing from the plain into the Arkansas River brought sand, silt, and smaller clay particles to the river. When the Arkansas River flowed into the Mississippi River, it brought this load of sediment with it. All of this sediment slowly flowed down the Mississippi River to the Mississippi Delta and the ocean. The current stopped when the ocean was reached and the pebbles, sand, silt, and clay fell out at the delta. Long shore currents in the ocean also moved sediment to other places on the shore. Very fine particles in suspension were deposited farther out in the ocean.

Over time, this loading has caused the area of deposition along the shore and just offshore, to sink at about the same rate as deposition replenished the shore. There is a time lag on the sinking so that when the loading stops, the sinking continues. The current sinking is responding to loading that occurred centuries ago. This sinking is called isostatic readjustment. This process works both ways; in Glacier Bay, Alaska, just the opposite is taking place. There is an isostatic rebound or rise of about half an inch per year occurring because of the melting of the ice in the valley over 100 years ago. With the unloading of the ice, the land surface rises. The area around the Great Lakes is still rising at a very slow rate in response to the melting of the last great ice sheet over 11,000 years ago.

Now let us look at what has happened since humans started to make an impact on the land. Throughout the drainage basins all over the United States, there have been over 75,000 dams built. The white settlers' damming started with a water mill at Plymouth, Massachusetts, in 1637 and has been going on ever since. Gullies have been dammed for stock ponds. In the Flint Hills of Kansas, this is a major source of water for cattle in the pastures. Streams have been dammed for fishing lakes. Many counties in Kansas have fishing lakes built by the Civilian Conservation Corps during the 1930s Depression.

Rivers are dammed to create a source for city water systems. Emporia, Kansas, has dammed two rivers, the Neosho and Cottonwood Rivers. The Neosho has better water and the Cottonwood was the backup source. Now, however, when the Neosho runs low, water is released from the large Council Grove Lake about twenty-five miles upstream on the Neosho River. Many large dams built for cities' water supplies also provide flood control and recreation. In the arid parts of the country dam projects have been built to provide irrigation water for agriculture. Most of the large dams are designed to produce power. There are dams and locks on the larger rivers to allow increased water transportation.

While the US was building dams, we were also pumping out large amounts of ground water. In many small towns, every residence and business had a well to pump their water needs. Cities like Wichita, Kansas, got all their city water from wells until about forty years ago. Irrigation supplied by wells has used up large amounts of ground water. All of these wells have caused a drop in the water table so that most rivers and streams no longer get water from springs. With the drop in the water table and the damming of the rivers and streams, there has been an extreme drop in the water flow in the rivers. One other thing has occurred, and this is the expansion of cultivation and the resulting increase in soil erosion by water. This has put more silt and fine clay particles into the rivers and streams.

Water and sediment flowing into stock ponds and fishing lakes usually stay there and go no farther. In the larger lakes behind dams, water and sediment come into the lake on the upstream side. Since the current stops, the sediment falls out on the upstream side, forming a small delta in the lake. Water, with only minor amounts of sediment in suspension, flows over the dam or spillway at the downstream end of the lake. This has happened in hundreds of lakes on each river system so that no sand or larger material has reached the shoreline for a number of years. About the only thing that has gotten past the dams are colloidal sized particles and very fine clay particles in suspension. These are carried far out into the oceans and do not drop out near the shore.

Without fresh sediment to maintain and build up the shores, they sink. The Mississippi Delta has sunk a great deal and continues to do so. The waves and storms have washed sediments that were at or near the shore far out in the ocean. In some cases, sediments from shore have been washed far inland during hurricanes with large storm surges. This is one of the reasons that some of the beaches are disappearing. Also, this is one of the reasons why some people think the ocean is rising. Since I live in the middle of the continent, this just has scientific interest to me. If I had a beach house on a disappearing beach, I might be a little upset about those landlubbers on solid ground with all their dams, keeping me from getting my fair share of sand.

This is not just an American problem; this is a worldwide problem. There are about 800,000 dams throughout the world. By world standards, the United States is not one of the major dam builders. Of the fifteen highest dams in the world, none is in the United States. None of the world's fifteen largest-capacity reservoirs behind dams is in the United States. Of the world's fifteen largest-capacity hydro plants, not one is in the United States.

The high Aswan Dam on the Nile River in Egypt created Lake Nasser, which is about 200 miles long. In our discussion of ancient irrigation in Egypt on the Nile, we talked about the fertile silt that came with the Nile River water to provide nutrients for the plants. The Nile silt is now deposited in the south end of Lake Nasser in the country of Sudan. The irrigation projects on the Nile now get nutrients from chemical fertilizers.

The sinking shorelines from lack of sediment occur all over the world. This is not a problem that will go away. It will get worse every year, putting coastal communities and structures in jeopardy. There is no way to get the sediment from the deltas in the lakes to the coast.

Plant and Animal Changes with Increased Atmospheric Carbon Dioxide

There have been many plant and animal changes in the last 150 years during the atmospheric carbon dioxide increase and the increase in human population. Not much attention is given to plant conservation, but many native plants have been plowed under and lost to deforestation. These were replaced by plants that could be utilized more effectively by people for food, clothing, and medicine. We ought to make a concerted effort to retain the varieties of plants that were in place before humans showed up. The interrelationships that link the native plants and animals are not well understood; in other words, we don't even know what we are losing. Plant diversity is a valuable resource for the other living things on earth, particularly man. With the increase in atmospheric carbon dioxide over the last 150 years, these relationships have changed and continue to change. Some animals, such as armadillos, have moved northward. They may have been following the movement of a plant community.

Populating the United States with farmers and ranchers took its toll on the wildlife in several ways. Early on, it became obvious that it would be pretty tough to raise a decent crop when herds of buffalo could come walking through the fields, trampling down whatever was trying to grow. Also, out on the range, the wildlife ate the same grass that the ranchers wanted the cattle to eat, so the wildlife was deemed expendable. It took time for a farmer to become established and grow his own food. He survived this lean time by reverting to the hunter-gatherer mode, living off the land and the wildlife. After the farmers were established, they still hunted to bring a little diversity to their diet.

In the 1930s, people started migrating away from the farms; this continued after World War II. At the same time, the wildlife conservation effort became more effective and the wildlife started to recover. In the 1960s, television and sporting events diverted interest away from hunting. Also, supermarkets provided a large diversity of foods with no effort so that hunting was no longer necessary to provide variety in the diet. The atmospheric carbon dioxide increase enhanced plant growth and wild animals had more to eat. This brought increases in the animal populations.

The increase in some wildlife was dramatic. In 1950, the deer population in Kansas was estimated at about 15,000. By 1990, the deer population had increased to 400,000. This may be greater than anytime in the past. Before the pipeline was built from Prudhoe Bay to Valdez, Alaska, some people were worried about the impact on the caribou herd. To check the effect, the caribou in the herd were counted. The population of the herd was counted again twenty years after the pipeline was completed, and the caribou population had doubled. The cause of the increase probably had nothing to do with the pipeline. It was more likely caused by the increase in atmospheric carbon dioxide, which allowed the vegetation to grow faster and the herd to double from extra food. Most animal species whose recovery has been encouraged have had a remarkable success.

Attitude Changes with Increased Atmospheric Carbon Dioxide and Population

Exploration of the earth was still a major function 150 years ago. To be successful at exploration requires a non-threatening attitude of interest and discovery, so that any interaction with the natives is friendly. The initial exploration in Australia was intended to see what was there. Africa and remote areas of Asia were explored in a more detailed way. The exploration of the Arctic and Antarctic came in the early 1900s. Most of the initial exploration was completed by World War I. The extent of the land and oceans had been defined. Much of the conquest that followed exploration was already accomplished and Europeans had established many new colonies.

Exploitation followed exploration and conquest. Exploitation requires a different mindset and attitude than exploration and conquest. Exploitation is the practical business of extracting the value out of what is exploited, such as land or mineral deposits.

The same difference between exploration and exploitation exists in the fossil fuel profession. The exploration geologist finds a new field; then the exploitation geologist determines where to drill development wells in the new field to get the oil out of the ground in the most efficient manner. (When the term "exploitation" became politically incorrect, they became "production" geologists.) The exploration person is a broad-brush type with an always-optimistic attitude. The exploitation person has to be detail-oriented and a realist, with a cautiously optimistic attitude.

The farmers were trying to make a living off the land 150 years ago with little knowledge to exploit the land effectively. There was a lot of trial and error involved with farming. These people did whatever they could to get a crop harvested. The supply of land appeared to be inexhaustible and if the land gave out, they could move on.

The initial exploitation of fossil fuels was relatively small and localized and consisted mainly of coal and oil production. There was little or no thought about possible damage to the land or water supplies; getting the fossil fuels out of the ground was the only concern. The attitude was to deplete the fossil fuel here and find something better somewhere else.

Many people in the mineral exploitation business started out looking for gold. When their search was unsuccessful, they had to work for someone else. Often the search for gold turned up other economic minerals and people mined whatever could be sold. By looking at the mounds of mine tailings and the prospect pits in the Rocky Mountains, it is evident that the miners did not think about the future of the land. Their sole concern was getting the minerals out of the ground.

Transportation was pretty slow in wagons, on horseback, or in the early trains and the continent seemed limitless. A sea voyage from the east coast to the west coast had the feel of going to a different world. The Transcontinental Railroad changed that. It had a unifying effect on the country and defined the limits of the continent. People's attitudes changed as they viewed themselves as part of the country. By the early 1900s, much of the country was populated and doing fairly well.

Throughout history, people have been travelers, but World War I made people take a global view of the world and view other people in different ways. Soon thereafter, the Dust Bowl of the 1930s brought Americans a new attitude toward the land, as well. It became obvious that farming had to be done in a way to protect and improve the land. With World War II, we had people all over the globe, and the world seemed much smaller. People returning from overseas had a new perspective on the people and the world around us.

A few decades later, pictures of the earth from the moon altered many attitudes towards the earth. There was the realization that we all needed to work toward maintaining or improving the environment. The problem is that no one has that great book of knowledge that tells us what needs to be done. We have about 4,700,000,000 years of geologic history to draw on. A fair amount of data has been collected in the last thirty-five years that gives us a snapshot of one point in time. However, politics and economic considerations muddy the waters.

The global environment includes land, water, atmosphere, ice, outer space, plants, and animals. It is the habitat for living things. The earth, its resources, and its habitat are pretty much finite. Except for one small point on the big island of Hawaii and possibly a few other places of volcanic activity building land, there is no new land being created.

Most of the environment does not belong to anybody or to any nations. The oceans are not controlled by any single country, and the atmosphere is not controlled by anybody. Antarctica has lots of visitors, but no owners. All of these portions of the world are covered by loose agreements between nations. In effect, there is no one responsible and no one to blame for environmental changes in these areas. In these areas, it has to be a global concern and responsibility which is agreed to by the global community. This means a compromise negotiated by politicians, with science being a secondary consideration. Environmental attitudes are varied, but they will play a large role in mankind's future.

Chapter 13. The Way We Are and Where We are Heading

All that has been written so far has to do with the history of the Ice Age and more recent history. We focused on changes in atmospheric carbon dioxide and how these changes affected plants, animals, and humans. From now on, we will be looking at mankind's current situation and some of the things that can be expected in the future.

Where We Are Today

We are now in the later stages of the interglacial period. If the Little Ice Age from AD 1450 to 1850 was a random event, then there could be another 5,000 to 6,000 years of the interglacial period. However, if the Little Ice Age was the first blast of ice in the accumulation phase, then it's going to get very cold much quicker. The warm spell we are in now appears to be the best point in the glacial cycle for life development. We are fortunate to be living in the best of times, from a glacial point of view.

The world population of about 6,500,000,000 people is close to being in balance with the current food production. Some places have severe food shortages, which result in recurring famine. In other places, food surpluses are not always effectively utilized because of political or economic disruption. The areas with food shortages just about balance with the areas of surplus food and productive capacity.

However, there are some disturbing aspects of the current food situation. World production of wild fish may have peaked. Almost all of the fertile land available in the world is under cultivation, and some of this land is already being lost to expanding cities. Also, fertile land is being damaged and removed by wind and water erosion. The increase in food production in the last hundred years has come mainly from more efficient crop yields, most of which can be attributed to the high levels of atmospheric carbon dioxide. At least the atmospheric carbon dioxide continues to increase.

The use of energy has increased at a rapid pace during the last 150 years. Most of the energy we use now comes from fossil fuels and this will continue for some time. Adequate supplies of fresh water are an absolute necessity for plants, animals, and humans. Most freshwater resources are used by agriculture, cities, and industry.

Most of the economies of the world are expanding and that is the goal of all countries. The economies of China and India, which hold about one-third of the world's population, have experienced rapid growth and that is expected to continue. This expansion requires greater use of land, natural resources, and fossil fuels. For the most part, these are non-renewable resources.

As the human population has increased, it has encroached on the habitats of wild plants and animals. In many areas, humans have replaced wild plants and animals with domestic plants and animals for food and fiber. Much of the native grasslands of the United States' Great Plains have been replaced by wheat, corn, other grain crops, and cotton. The vast herds of buffalo, elk, deer, and pronghorns have been replaced by cattle, sheep, and hogs. The population of animals in the wild has decreased significantly in most areas, particularly in the last 200 years. Some animals, such as the whooping crane and California condor, have been close to extinction; the viable populations of these animals are managed by humans.

We appear to be close to a turning point in the human domination of the earth. The population growth and economic expansion that has occurred for the last 150 years is clearly not sustainable for very far into the future. The details in each area give us a better picture.

Human Population

We live on a finite earth with certain physical limitations. Water in the form of oceans, glaciers, lakes, inland seas, and rivers cover about 75% of the surface of the earth. The remaining 25% has dry land on the surface and covers about 50,000,000 square miles. This area has to support all land plants and animals, including humans. With over 6,500,000,000 people on earth, this means that there are over 130 people for every square mile of land surface.

If people were distributed equally throughout the land surface, each individual man, woman, and child would have five acres of land to supply all their needs. This land must also supply the needs of the other animals and plants on earth. I certainly would hope that I would not have to share my five acres with lions and tigers and bears — oh my! I have always loved the desert southwest, but I am not sure I could survive on five acres of prickly pear, rattlesnakes, and other things that stick, sting, or bite. I would hope that my five acres was good bottom land that did not flood. It would be tough living if I got a mountaintop or the frozen tundra.

We fairly well have a closed system on the land portion of the earth. If the wild animal population increases, animals encroach on the human population. Deer and rabbits eat the flowers and vegetable gardens, and on very rare occasions, tigers have been known to kill people. Humans, in turn, encroach on the space needed by other animals and by plants, and routinely kill them. Population growth and development both have a large effect on the world around us now and in the near future.

The human population growth is the greatest in the lesser-developed countries that are located mainly near the equator. Most of these countries have a very young population. The people form families at a young age and have a high birth rate.

The medical profession has made a great contribution to mankind and has done an excellent job of increasing the infant survival rate and in lengthening the life span of the people. The World Health Organizations have brought modern medicine to the lesser-developed nations. Consequently, the population of the lesser-developed nations has increased in several ways. There are more live births and more of the infants live to be adults. Also, the adult population lives longer.

Some religions that reject birth control and promote large families are firmly established in many lesser-developed countries. If these religions are in countries that can't feed themselves, then they damage the whole country.

Most of the developed countries are located in the mid to high latitudes of the world. They are in areas where much of the soil is rich from the glacial processes and the formation of loess. The food for the population is produced by a small number of farmers. The rest of the population can do other jobs, often in big cities. The developed countries require more resources to maintain or increase their standard of living. They need minerals, metals, and building materials. More land and materials are needed for cities, roads, airports, and bigger dwellings. Developed countries require fossil fuel in the form of coal, oil, and natural gas to provide warmth, mobility, and petrochemicals. Material success is a major goal in the developed countries and is becoming a goal in all countries.

Since most developed countries are very efficient in agriculture, many have a surplus of food. Some of this food goes to the lesser-developed countries. Many of the lesser-developed countries have a surplus of people and some of these people end up in the developed countries which have more stable populations.

The growth rate of world population on a percentage basis has been declining for the past forty years, and this is expected to continue into the future. Still, the total world population is increasing; it is expected to top out in about fifty years at 9,000,000,000 to 10,000,000,000 people. From there, we will have a gradual decline.

For the last 12,000 years, the limiting factor on vegetation growth, food, and human population has been atmospheric carbon dioxide. From now until the fossil fuels are all burned and the atmospheric carbon dioxide declines, the limiting factors may be land, water, nutrients, and personal choice.

Energy

The modern world runs on energy, and the future of the economies of the world is dependent on adequate energy supplies at reasonable costs. There are a number of sources of energy, but the major sources have been developed over the last 150 years. The energy consumption in the United States in 2004 was as follows: petroleum 40.1%; natural gas 23%; coal 22.4%; nuclear electric power 8.2%; wood, waste, and alcohol 2.85%; hydroelectric 2.7%; geothermal 0.3%; wind energy 0.14%; and solar 0.14%.

The fossil fuels, petroleum, natural gas, and coal, account for over 85% of the energy consumed in the United States, and the world consumption of energy is similar. This is truly the age of fossil fuels. Some politicians would like to change this by passing laws to cut down fossil fuel usage; however, there is no viable alternative. We either use fossil fuels or do without.

We will look at where we are today in usage and development of a number of energy sources. The fossil fuels will be reviewed first.

Oil

The world gets more energy from oil than from any other source. Many people think of oil as black gooey stuff that comes out of the ground, from which oil companies make gasoline and messes. Actually, oil is a very complex mixture of stuff and is different in composition at each location where it is found. Some oil is almost clear and has the consistency of gasoline. Some oil is solid at room temperature and has to be heated to flow. There is even some oil that is black gooey stuff and looks like a mess waiting to happen.

Chapter 13. The Way We Are And Where We Are Heading

The products of the oil industry are intimately linked to the lives of all of us. These products help to provide with light, warmth, and mobility. They also provide lubrication for all the machines. Petrochemicals are the basis for most plastics, a number of fabrics, and many other items of everyday use. Modern farming and the food supply are dependent on the fuel used to power the farm implements. The delivery of all the goods we receive is fueled by oil products.

The style of life that we have become accustomed to is dependent upon plentiful supplies of oil and gas. During times when the supplies of oil are not plentiful and cheap, our lives are disrupted and we start growling.

One of the best ways to see where we are today in the oil business is to compare it with the past. Today's oil production and consumption worldwide is just about equal to the world's productive capacity. The last time these two were about in balance was right after World War II, about sixty years ago. At that time, the United States had over twenty major integrated oil companies and United States companies owned about 70% of the world's reserves. All of these companies had one motivation and that was to get bigger. To do this, they spent much of their cash flow on drilling and other capital expenditures to find and produce more oil.

Today, the United States has fewer than ten major integrated oil companies and United States oil companies own less than 10% of the world's reserves. Over 75% of the world's reserves are owned by producing countries. Most of their cash flow is spent to run their governments. Not much goes into finding and developing new oil reserves.

Proved reserves today worldwide are less than they were in 1980. If we use the strict definition of proved reserves that was used in the 1970s, today's reserves are much less than they were in 1980. Since 1980, production and consumption have increased by 40% worldwide. The energy use in China and India is increasing dramatically. If these trends continue, there will be a shortage of energy or a rapid increase in price, or both. When OPEC got pricing power in the 1970s, the members got feisty. Iran's nuclear stance suggests that history is repeating itself.

Oil is available in large quantities in only a very few unique places in the world. Finding and developing oil has a very long lead time. We need to start now to find oil for the next generation to use. When we choose not to develop the sweet spots in the United States for any reason, then we become more dependent on potentially unstable sources of supply from around the world. For instance, by not looking at the 8% of the Arctic National Wildlife Refuge in Alaska that has big oil potential, or the Outer Continental Shelf, we are drastically shortening the time until depletion of all United States oil resources.

In times of shortage, incentives to increase domestic production should reward only successful efforts, not just activity. A great honking and flapping of wings won't cut it; we have to fly. Although it has been a politically dirty word, the depletion allowance was the most effective incentive, because the reward came only when the oil was in the tank or the natural gas was in the pipeline. The depletion allowance does the same function for the extractive industries as depreciation does for manufacturing. There are very large capital costs in exploring for oil and gas. About ten exploration prospects are drilled for every one that finds commercial oil or gas. The depletion allowance allows the sale of some hydrocarbons tax free to keep from paying tax on the recovery of the original capital expenditure.

The trends occurring now suggest that peak production of conventional oil can occur anytime within the next fifty years. There will be a decline in conventional oil production from that point. How fast the production will decline has yet to be determined.

Tar Sands

Tar sands is a loose term for a sandstone containing petroleum where the lighter oils have escaped, leaving a tar residue. The difference between a tar sand and a heavy crude oil reservoir is sometimes fuzzy, and in the United States, most go by the latter term. The two largest tar sand reservoirs are the Athabasca Tar Sands in northern Alberta, Canada, and the Orinoco in Venezuela.

The Athabasca Tar Sands are about 150 miles long and 75 miles wide. The Tar Sands are exposed at the surface in some places. The depth of burial of the Tar Sands ranges from zero to 1,000 feet. The Athabasca Tar Sands have about 696,000,000,000 barrels of very heavy oil in place. In the late 1960s, the Great Canadian Oil Sands Company started mining the Tar Sands on the surface and built an oil extraction plant to process the tar sands. The tar is upgraded in the process of removing it from the sand, and the product is similar to diesel fuel, but requires further refining. The Great Canadian Oil Sand Company was at that time a wholly owned subsidiary of Sun Oil Company.

In 1970, I was transferred to Calgary as a petroleum geologist for Sun Oil Company. While I was in Calgary, the boss wanted to know where to buy more reserves for the plant, and the total recoverable reserves for the Tar Sands. Our reserve evaluation gives some insight on the exploitation of the deposit. To find the recoverable reserves, we looked at the Athabasca Tar Sands in four parts: tar sands exposed at the surface; buried tar sands with less than 150 feet of rock over them; buried tar sands with 150 to 600 feet of rock over them; and buried tar sands with 600 to 1000 feet of rock over them.

The initial mining of the Athabasca deposit was where the tar sands were exposed at the surface. The total recoverable oil was figured on this volume of tar sands. Tar sand beds buried by less than 150 feet of overlying rock were considered to be recoverable by strip mining sometime in the future. The price of oil would have to be much higher for strip mining this deep to be economic.

Where the tar sands are buried from 600 feet to 1,000 feet, the oil would have to be recovered from wells using thermal methods. The thermal methods include injecting steam to heat up the oil and the reservoir to make the oil flow. Downhole combustion is another way to get heat in the reservoir. Today most areas use steam injection. The areas in Athabasca with 150 to 600 feet of rock over the tar sands were considered unrecoverable. It was believed to be too deep for strip mining. It was believed to be too shallow for thermal recovery. When the well was under pressure from steam or air injection, the pressure would bleed to the surface. Our estimate of total recoverable reserves from Athabasca was 150,000,000,000 barrels of oil.

In Alberta, there are two other areas of tar sands that have been producing oil using thermal methods. These are at Peace River and Cold Lake. The tar sand reserves in all of Alberta are currently estimated at 175,000,000,000 barrels of oil. These will take a very long time to produce.

Oil Shale

Oil shale is a broad term for shale or shale with lime which contains large amounts of kerogen. Kerogen is a solid hydrocarbon which, with heat and pressure, becomes liquid oil. Destructive distillation of the oil shale produces shale oil and gas. The kerogen starts to liquefy at about 900 degrees Fahrenheit. The higher grade oil shale yields about twenty-five to thirty-five gallons per ton of shale.

The United States oil shale resource base is about equal to the proved crude oil reserves of the world. The world oil shale resources are about double what is in the United States. Oil from oil shale has been produced intermittently since the 1800s. It is currently being produced in China, Brazil, and Estonia.

The oil shale resource base is a measure of shale oil that is technically recoverable. Only shale oil that is economically recoverable will be produced some time in the future. When the production will occur and how much will be economically produced is not known.

Natural Gas

Natural gas, like oil, is a depleting non-renewable resource. The natural gas producing industry is interrelated with the oil production industry. Natural gas is produced as a by-product in almost all oil wells, in quantities ranging from a small percentage to gas being the most important product. Conversely, many gas wells have no oil produced with them.

In exploring for gas or oil, you often cannot tell the content of the reservoir before you drill the well. You may find oil or gas or a mixture of the two, or an oil reservoir with a gas cap above it. Most of the time all that is found is saltwater. Sometimes you don't even find any porous rock to provide a reservoir.

Sometimes gas wells produce gas liquids with the gas. Carbon dioxide, nitrogen, and hydrogen sulfide can also be found in some gas wells. This gas goes to a gas plant where the gas liquids are removed. The nitrogen or carbon dioxide is removed; if the amount is small, they may just be diluted before going into the consumer pipeline. Hydrogen sulfide is a deadly gas and has to be removed before it gets into the pipeline.

I worked on some 20,000-foot wells in West Texas that were drilled to the Fusselman Formation. The Fusselman gas had 4% hydrogen sulfide. The government regulations required that while drilling the well, we had to have air packs and a flare gun at the well. In the event of a major blowout in the Fusselman, the air packs were to get the workers off the rig. The flare gun was to set the rig on fire to prevent poison gas from drifting into a populated area. The Fusselman had normal pressure for 20,000 feet in depth so there was almost no chance of a blowout.

There is a lot of confusion about how much commercial natural gas there is in North America. People in the natural gas industry who find the gas, get it out of the ground, and transport it to the end user, you and me, talk in terms of proved reserves. Government agencies, policy makers, and various other broad-brush types talk about natural gas in terms of technically recoverable resources. The big difference is the people working in the field have to deal with the economics of the business. The broad-brush people have no concept of the economic realities.

In 2004, the proved reserves in the United States were 189 trillion cubic feet of gas. This was about 3% of the proved reserves of the world.

The United States energy information agency in 2002 put the United States natural gas technically recoverable resources at 1,190 trillion cubic feet of gas. The unproved reserves are six times the proved reserves. After thirty years in the oil and gas business, I would suggest that it is a lot easier to say there is an abundance of gas than to find that gas in reservoirs that can be economically developed and produced.

One of the biggest problems is getting leases on the land that has potential for major gas production. Much of this land is federal land and is withheld for many stated reasons. Many organizations use environmental protection as an excuse to lock up the land so that no drilling can occur. Congressional action is often needed to open up land for leasing and this almost never happens. If the country is going to use natural gas, many hard choices need to be made. If land is not available for drilling, then many gas producing companies will go out of business.

In the United States, all gas has to be moved in pipelines. If gas is found in remote areas, then a lot of development wells have to be drilled before the cost of a pipeline is justified. Many small or medium-size gas deposits are not brought to market because the pipeline would cost too much. This is a big problem on the world markets. Many large gas deposits have not been developed because there is not a buyer for the gas, and no way to transport it.

In the United States, we import gas in pipelines from Canada and Mexico. We also import liquid natural gas carried in liquid natural gas tankers. This requires a plant to liquefy the gas in the exporting country and a terminal and liquid natural gas storage in the United States. The liquid natural gas tankers and the facilities in both countries are all very expensive.

How long we rely on natural gas in the future will depend on the will of our people and the size of their pocketbooks.

Methane Clathrates

There has been speculation about the use of methane clathrates as a source of natural gas in the distant future. Methane clathrate is a funky form of water ice with large volumes of methane confined in its crystal structure. It is sometimes called burning ice. Under normal pressure, methane clathrates are stable up to about 65 degrees Fahrenheit. It is currently believed that there is considerably more methane in methane clathrates than in all natural gas deposits.

Methane clathrates are found in shallow marine areas, deep sedimentary formations, outcrops on the sea floor and in Arctic areas. At this time, there is no known way to get the methane out of most of these deposits.

Long before I ever heard of methane clathrates, I suspect I found a reservoir containing them in a place not currently recognized. In the 1970s, I was the well site geologist on a gas well drilled about four miles deep to the Fusselman Formation in the Moore-Hooper field in West Texas. We had to drill through an over pressured zone at about three miles deep, which was the Atoka limestone. We took measurements with electrical, sonic, and radioactive tools, which indicated that the Atoka could produce natural gas. Ten years later, after the Fusselman Gas Well was depleted, this well was recompleted in the Atoka.

The typical oil or gas wells are drilled in formations that were originally deposited on the bottom of the sea. Oil and gas were formed and moved into the formation later. When the oil and gas moved into the formation, most of the ancient sea water was pushed out. However, a small film of salt water remained on the rocks of the formation. The salt content of this formation water can be measured. The Atoka Formation water was measured by electrical means and had a salt content of about 80,000 parts per million.

The Atoka Formation in this well had a temperature of about 185 degrees Fahrenheit and a formation pressure of about 15,000 pounds per square inch. After the Atoka Formation was completed and had produced for awhile, the production rate was 1,500,000 cubic feet of gas and 600 barrels of fresh water per day. At that time, we had no clue where the fresh water was coming from, as such wells generally produce salt water. It was many years later when I read about methane clathrates that I decided that this must be the source of the fresh water. These methane clathrates may have been sitting in the Pennsylvanian Atoka Formation for 300,000,000 years.

In the past, when the oil men found these very high pressure zones, sometimes there was a blowout and the rig burned down. This would be a tough way to find methane clathrates. How to exploit them once they are located is another challenge.

Coal

Coal is a sedimentary rock derived from vegetable debris, which is formed by time, heat, and pressure. It is a black carbonaceous material that can be burned for energy. Coal is a depleting non-renewable resource. The United States has about 25%s of the world's coal reserves. By some estimates, these coal reserves are equal to the Middle Eastern Oil Reserves in energy equivalents. At the current United States production rate, reserves of coal could last 250 years. In the United States, coal provides 22.4% of the total energy and half the energy for electric power generation.

Coal is mined from the Colorado Plateau to the Appalachian Basin and from the northern Great Plains to the Gulf Coast. Much of the coal is hauled by unit trains where the whole train load goes to one destination. Having the coal supply close to the power plant is very economical.

The large western strip mines produce a low sulfur coal which does not have the environmental problems that many of the eastern coals have. They also are very efficient and competitive in price with most energy sources. Much of the eastern coal comes from underground mines which are more expensive to operate. For this reason, much of the eastern coal is not as competitive in price.

Chapter 13. The Way We Are And Where We Are Heading

Because of the high sulfur content of much of the eastern coal, the end user often must install scrubbers to remove the sulfur. Their main advantage is that they are closer to the big markets, so transportation costs are lower.

Outside the United States, major coal reserves are found in Russia, China, Australia, Germany, India, and South Africa. Coal will probably be a contributor to world energy use long after the other fossil fuels are depleted. It is the largest energy resource available in the United States.

Nuclear

Uranium, which is the fuel for most nuclear reactors, is a depleting non-renewable resource. If we should decide to build breeder reactors, we would have a renewable resource because they create more fuel than they use. In the past fifteen years, some of the uranium used in power generation came from Russian stockpiles, dismantled nuclear weapons, and reprocessing of spent fuel rods.

Nuclear electric power accounts for 8.2% of all energy used in the United States and provides about 17% of the electrical power. Worldwide, about 17% of the electricity is produced by nuclear power reactors.

The utilities in the United States have not ordered a new nuclear plant since 1978. The last nuclear reactor to be built started generating electricity in 1997. There is renewed interest in the United States in the new, more efficient, and safer reactors. Work should begin on more nuclear power within the next decade.

Worldwide interest in nuclear power is alive and well. China and India are both very interested in nuclear power to supply electricity to their rapidly growing economies. Several different types of reactors are being researched. There has been a reduction in nuclear power construction, but that will change in the coming decades as few viable alternatives exist.

In 1958, a review of nuclear fusion research said that the first nuclear fusion reactor would be built by 1985. In 1995, a report suggested that the first nuclear fusion reactor would be built in 2040. If it is ever built and produces electricity economically, it may make a major contribution to man's energy needs.

Hydroelectric

Hydroelectric power is mostly renewable. Hydroelectric facilities in the United States provide about 2.7% of the total electricity. Because of high initial cost, lack of sites, and high environmental land damage, future hydroelectrical development in the United States is limited.

Worldwide hydropower provides 19% of the world's electricity supply. In many places, land costs are much less so there are potential sites for dams and hydroelectric power plants. Some of these sites will be developed in the future.

Geothermal

Geothermal energy may be depleting locally but is renewable on a global scale. What this means is that at any one place you can often produce more heat than is being replaced naturally. When a geothermal source is found, it is often difficult to determine how long it will last until extensive historical data on production, temperature, and pressure have been collected. However, a depleted geothermal source may be renewed some time in the future as the earth continues to give off heat.

Geothermal energy in the United States produces 0.3% of our energy. It is used to generate electricity at the Geysers Field in northern California. This field produces 2,000 megawatts of electrical energy. This field is located relatively close to the end users, avoiding the large transmission costs that would be involved in working with a more remote location. There is also a geothermal plant in Hawaii.

Worldwide, geothermal heat is used in Iceland, Italy, and New Zealand for power generation and heating. In addition, many areas throughout the world use geothermal energy from hot springs for thermal baths. The earth's geothermal potential is limited to volcanic areas.

Alcohol, Wood, and Waste

Alcohol, wood, and waste produce about 2.85% of the energy in the United States. Waste is renewable; we can always generate lots of waste. Power generation from waste is far less efficient and cost effective than using fossil fuels, but the payoff comes when the alternate disposal costs are considered. There is a shortage of land for landfills. The supply of waste seems to be unlimited for the near future. Big problems for waste-to-energy plants include finding sites to build on and obtaining state licensing.

In the United States, wood is used for fuel in fireplaces and wood stoves. Some factories burn wood packaging. Wood waste from the lumber industry is used for fuel. Worldwide, wood is reckoned to account for about 6% of the energy used, but this is hard to measure. It would be difficult to increase wood use very much in the non-forest areas.

Ethanol production from corn has increased rapidly in the United States. Most of the ethanol is used as a gas additive, which is mandated by the government. There are several problems with ethanol. Nobody seems to have hard data on how many BTUs of energy it takes to make a gallon of ethanol from corn. Therefore, we cannot compare that with the BTUs of energy in a gallon of ethanol. Does it cost more than it is worth? In addition, each bushel of corn used to make ethanol is a bushel of corn not eaten by a human or domestic animal. Brazil and several other tropical countries make ethanol from sugar cane, which produces significantly more energy for the input.

Wind

Wind is a renewable but unpredictable energy resource. There are many wind power generating facilities in the United States and the number is growing. Wind power generation currently produces about 0.14% of the United States' energy usage. Since the late 1970s, the utility power grids have been required to buy wind generated power from whoever produces it.

Electricity in the power grid cannot be stored. It has to be used when it is generated, or it is lost. Wind power is intermittent. In Kansas, the wind blows a lot in February and not much in August. The winds are low at dawn and dusk when household electrical usage is high. The wind speeds are higher during the day. In this part of the country, wind produces commercial energy only about one-third of the time. There are a number of wind farms in Kansas currently producing or under construction. None of these are owned by the utility that uses the power. In order for the utility to effectively use the wind power, it must be paired with a fast start peaking unit to provide power when the wind isn't blowing. The utility company is currently building a natural gas-fired fast start electrical peaking unit north of Emporia, Kansas. This may not be the best power source, but government regulations make it necessary.

Worldwide wind has been a source of energy for hundreds of years, and the use of wind will probably continue to grow.

Solar

Solar energy is a renewable, but intermittent, resource. The most important and least quantified use of solar energy is passive use for light and heat. Solar energy and conservation are very closely connected. The use of solar energy is just a matter of trapping and conserving the energy from the sun. A hundred years ago, housing was designed to utilize heat from the sun in winter when the sun came in at low angles, while the porches provided shade in the summer when the sun was high. However, they did not have effective ways to store the heat for later use. There was little or no insulation to stop the heat loss. Ironically, when central heat became common, some of the more efficient house designs that utilized solar heat were abandoned. We have now gone back to using some of these older designs, but we also have improved on them and have added temperature-stabilizing features like insulation and multi-paned and coated windows.

Computer-controlled management of solar heat is very productive in large glass box buildings. In these systems, warm air on the sunny side of the building is circulated to the inner portions of the building and to the shady side of the building to maintain a constant temperature throughout the building. Also, solar heat can be used as a supplemental water-heating source.

Electricity produced by photovoltaic cells costs more than fossil fuel electricity, but this is applicable for remote locations. In remote locations, the cost of transmission systems greatly increases the cost of conventional electricity. This energy source should also grow as the price difference narrows.

The biggest drawback to solar energy is that it is intermittent. Energy storage is not very efficient, so this is an area which will need more research.

Solar use will become more important as the fossil fuels are depleted.

Probably other energy sources unknown today will be developed and become important in the future.

Carbon Dioxide

Carbon dioxide in the atmosphere is increasing at a rapid rate. There is no system in place to verify exactly how fast it is rising; the only data ever quoted comes from the Mauna Loa Volcano, which gives off large amounts of carbon dioxide. It is twenty miles from the Kilauea Volcano, which has been the biggest source of atmospheric carbon dioxide on earth since it started erupting in 1983.

The increase in atmospheric carbon dioxide mainly comes from two sources: burning of fossil fuels and from the ocean. The atmospheric carbon dioxide is a very small part of the carbon dioxide in the ocean–atmospheric system. There is about sixty times as much carbon dioxide in the oceans as there is in the atmosphere.

If the temperatures of the surface of the oceans are stable for long periods, then the carbon dioxide in the ocean–atmospheric system can reach equilibrium. This means the amount of carbon dioxide released from the warm water of the oceans near the equator equals the amount of carbon dioxide absorbed by the cold ocean surface near the Poles. Carbon dioxide in the ocean–atmosphere system was in equilibrium throughout most of the interglacial period until the Little Ice Age. The surface of the oceans has warmed since the end of the Little Ice Age. The oceans would be releasing carbon dioxide if the burning of fossil fuels had not drastically upset the equilibrium.

For the last hundred years, the burning of fossil fuels has been putting significant and increasing volumes of carbon dioxide into the atmosphere. This is occurring much faster than the oceans can absorb the excess carbon dioxide.

The increase in human population is expected to peak in about fifty years. Fossil fuel usage should peak sometime in the next fifty years and be relatively insignificant in about 300 years. This would suggest that the atmospheric carbon dioxide would peak in 50 to 100 years and decline from there, and decline rapidly in 300 years.

Our crystal ball is always cloudy and many things occur that cannot be predicted to alter the carbon dioxide release. Wars, political, and religious events in oil-producing areas always muddy the waters. People often do not act the way one would anticipate. New technology can change the world. If the surface of the oceans cools, the carbon dioxide can be absorbed faster. As carbon dioxide builds up in the atmosphere, absorption in the oceans may occur faster.

When the carbon dioxide in the atmosphere peaks, the maximum level cannot be predicted because of the many variables. In the past 150 years, the carbon dioxide has increased by 80 to 100 parts per million. By projecting this increase for another 50 to 100 years, an addition of 200 parts per million would be a maximum to the peak. This would give a peak of less than 600 parts per million atmospheric carbon dioxide. This is a guess and it is probably high.

Land and Water

The land devoted to agriculture covers 35% of the land surface. Of that amount, 13% is under cultivation and 22% is pasture land for grazing domestic animals.

The 13% of the earth's land used to raise crops has to have enough water for the plants to grow, either from precipitation or irrigation. The soil has to be fine grained so the roots of the plants can penetrate it. There have to be nutrients either from the soil or from added fertilizer. The land is often more productive with the addition of herbicides and pesticides. The soil has to be deep enough to accept the plow. Most land has too many rocks to plow. Most land that can raise crops is currently being used for that purpose.

In the future, this agriculture acreage base will not increase and there are trends in place that will make it decrease. Human population is expected to increase by about 2,500,000,000 people in the next 45 years. These people will need cities, towns, and structures in which to live. Many of these will be in the lesser-developed countries. As pressures build to improve living standards worldwide, construction will continue to take land out of cultivation in the developed countries. Erosion still destroys and damages farm land. Inevitably, it seems, the land devoted to growing food will slowly become smaller.

Fresh water is used by humans for agriculture, industries, and households. The need for fresh water is greater than the supply in many areas of the world, and this is expected to affect many more areas in the future. The quality of the water in many areas is also at risk in the future.

Irrigation worldwide uses about 70% of the fresh water today. Irrigation in some desert areas supplies all the water to grow plants. In many areas, irrigation water is a supplement to precipitation in order to increase crop yields. About one-third of the food grown in the world today comes from irrigated fields. In the future, more efficient methods of irrigation will have to be implemented in many areas.

Household water use worldwide consumes about 15% of the water supply. This is for drinking, bathing, cooking, sanitation, cleaning, and watering grass and other plants. Some of this water is collected, treated, and reused by someone else.

Industrial users worldwide take about 15% of the water used. Hydroelectric plants use water as the power source. Other power plants use water for cooling. Oil refineries and chemical plants use water in chemical processes.

When there is a shortage of water, as on the Colorado River, and there is a conflict between agriculture, cities, and industrial uses, cities almost always win. Industrial users lose to cities but win over agriculture; farm irrigation almost always loses the water.

There are potential problems with the current water supply and in many areas, it will be difficult to supply future needs. Water comes from precipitation, surface water, or underground aquifers. Many aquifers are not readily accessible from the surface. Aquifers that are exploited can be run dry; when the water is gone, a new source of water must be found for future use.

The quality of the water can be a big problem. In dry areas, the salt (mineral) content of water often rises over time, and even in sources that are usable now the salt content may become too concentrated to use sometime in the future. Pollution in some areas makes the water dangerous to use. Water treatment plants are expensive and are not available in some lesser-developed areas.

World water supplies for the most part are dependent on climate and population. In the next fifty years, world water supplies will be short in areas with a dry climate and areas that have increasing population. In the bigger countries, water supply is a local issue, not a national issue. For instance, in the United States, the eastern part of the country has adequate water supplies while many of the western states have tight water supplies — which will get tighter in the next fifty years. In the future, getting and maintaining adequate water supplies will be the number one priority for many people.

Chapter 13. The Way We Are And Where We Are Heading

Food

For the last 150 years, food production has risen dramatically and much of this increase can be credited to increased atmospheric carbon dioxide. In some cases, the increase in food supply is greater than the increase in population. There were many major famines in India and China 50 to 60 years ago. Now, there are about three times as many people in India and China, and those countries not only feed their people but export some of the surplus.

Worldwide food production today is about in balance with the world population. Some areas have excess food which shows up in larger tummies or is exported. Other areas are short of food and have famines. If the world distribution of food were perfect, then everyone would have enough to eat. Of course, nothing is ever perfect.

The population of the world is expected to peak in approximately 50 years at about 40% more people than today or about 9,000,000,000 people. This will require 40% more food to feed everyone. Fossil fuel usage should peak some time in the next 50 years. Consequently, atmospheric carbon dioxide should continue to rise for at least 50 years, and it should peak between 50 and 100 years from now. This would suggest that the crop yields should rise from increased atmospheric carbon dioxide for at least another 50 years.

The rise in crop yields will partially be offset by a decrease in land under cultivation and less water for irrigation. Whereas the level of atmospheric carbon dioxide was the major limiting factor to food production in the past, access to land and water may be the major limiting factors to food production in the future.

Furthermore, with conventional oil production peaking sometime in the next fifty years, the cost of oil products will continue to go up in relation to other products. At higher prices, oil-based fertilizer, herbicides, and pesticides may become too expensive to use in farming.

The increase in population will probably exceed the increase in food production, as long as the population is rising. Famine in some areas will be with us in the future, as it has been in the past.

Where We are Heading

All the long-term projections on what will happen in the future have one thing in common. They are always wrong. Many events cannot be unforeseen and many relationships cannot be fully appreciated. Nevertheless, we will try to look into the future.

As we saw in studying the glacial cycle, the earth is a very dynamic place and change is the normal condition. Some people think we can stop change, or that we should freeze the environment of earth at one point in time. We cannot make time stand still, and conditions on earth are going to change whether we like it or not.

There will be more natural disasters affecting far more people than were affected in the past. The human population is rapidly growing, and many people live in areas subject to natural disasters. People living on the shorelines are subject to hurricanes and tsunamis; people living in cities in earthquake-prone areas are vulnerable in other ways. Volcanoes, floods and landslides pose a threat to many more people. There are a number of potential doomsday scenarios shown in "documentary" movies that could alter the course of civilization, but most of these are very rare.

Personal decisions by people all over the world can, however, alter the course. If more people choose smaller families, the population may peak out at a lower point and possibly sooner. If more of the people in less developed parts of the world wish to accumulate stuff and travel around like we do, the fossil fuels and other raw materials will be depleted faster.

Politics can alter the course. The people in power often make decisions that are bad for their country or the world around them. Most decisions are geared to accomplish short-term goals with no regard for what will happen in the future. Political choices often bring unanticipated results. Miscalculations and decisions intended to solve short-term problems can lead to a world at war. Political leaders today need to look at the long-term consequences of what they plan to do.

Economics plays a major role. Many projects are technically possible but will never be economically worthwhile. Most government studies never get past the "technically possible" part and leave the question of "how to pay for it" to someone else. How much fossil fuel will eventually be used depends on economic considerations. The amount of oil, gas, or coal in the ground is unimportant. The only important fact is how much can be produced at a price that someone is willing to pay.

One of the reasons that many studies avoid the economics question is that the economics of a scenario changes over time. Future demand and future supply cannot be determined. Consequently, it cannot be determined in advance whether a given scenario would be cost efficient. Still, they play a major role in determining where we are heading.

Chapter 13. The Way We Are And Where We Are Heading

Plants and Animals Going Forward

Two major dynamics affect wild plants and animals now and in the future. First, atmospheric carbon dioxide has increased dramatically and is expected to peak sometime in the next 50 to 100 years. This makes the plants grow faster, and grow in a larger area; this will continue until the atmospheric carbon dioxide starts to decline. The increased plant growth provides more food for animals to eat; in some areas wild animals have increased in population significantly. In some areas, they will continue to flourish in the near future.

The other dynamic is human intervention. Two major factors reduce the viability of wild plants and animals: an increase in human population and humans' ambition for a better life, including more material comforts. Both of these require more land and resources. This additional land means encroachment into areas occupied by wild plants and animals.

Wild plants that grew on the most fertile ground already have been replaced by domesticated plants in order to increase food and fiber production. Even on marginal lands, man is planting domesticated plants. In some parts of the world, wood is used as a fuel for warmth and cooking. Much land is being stripped of trees and brush to supply the fires of an expanding population. There are many environmental effects on the plant populations by the industrialized nations. These include plants being covered by airborne particulate matter and chemicals of many kinds, and water from contaminated rivers and underground aquifers affecting plants. Often the poorer the industrialized nation is, the greater the environmental damage. This is because environmental controls to industry are usually very costly, and survival is more important to these people than the environment.

While human encroachment is devastating to many animals, there are cases of wild animals that thrive in the presence of humans. The rat population of the world is probably expanding with the expanding human population. But on the most productive grazing areas and many marginal areas, most wild animals have been replaced by domesticated animals.

HISTORIC DATA AND FUTURE PROJECTIONS

Much of the view of the world around us has come from data that has been collected in the last hundred years. Where did this data come from and how was it processed?

Only two carbon dioxide detectors have been operating on a continuous basis since 1958. One is in Antarctica and one is in Hawaii on Mauna Loa Volcano. It seems that no raw atmospheric carbon dioxide data from Antarctica has been published in twenty years. At that time, the detector on Mauna Loa measured a much greater increase than the Antarctica detector. I have seen one graph with Mauna Loa carbon dioxide data and data called normalized Antarctica data. The normalization of the Antarctica data was to increase the values so it would be a direct lay down over the Mauna Loa graph.

The oceans are a great storehouse of energy, and they have a large influence on our weather. By looking at the effects of the Gulf Stream, the Alaskan current, the mechanics of hurricane generation, or the El Niño, we can see how important the oceans are to our atmosphere. Most of the moisture in the atmosphere comes from the oceans. The oceans and atmosphere are interdependent, with the oceans dominant in the relationship.

Much of the oceanographic data from the last hundred years has been collected at the margins of the ocean and on the surface of the oceans. The mean depth of the oceans is about 12,500 feet. Very little historical data and not all that much of today's data has been collected in the bottom 99% of the ocean's depth. The oceans cover 70% of the earth's surface, and over much of this area there is little incentive for extensive data collection. For example, the southern portion of the Pacific Ocean covers about one-fourth of the earth's surface, and very little oceanographic and atmospheric data has been collected in this entire area. This is because there is very little land and very few people there.

Historically, the bulk of the atmospheric data collection occurred on the land areas that fall between 20 degrees north latitude and 60 degrees north latitude. This is where most of the people lived who had an interest in predicting the weather. Not much long-term atmospheric data has been collected over the oceans, over the Poles, and in the southern hemisphere. The data collected in these areas is non-representative data, being found where it is available, not where you want it. Most of the theory and conclusions have come from analyzing the data where there is plenty of data to work with, particularly from Europe and the United States.

Much of the recent data comes from satellites. Satellites fall into three categories. First, there are the satellites that are launched from west to east and orbit the earth within about 30 degrees of the equator. These constitute most of the launches, because the rockets get a boost from the earth's spin. More payload can be put into orbit for the energy expended. Second, there are the fixed-position satellites, which orbit the earth at about 22,000 miles in altitude and stay over the same spot on earth at all times. The third category contains the polar-orbit satellites which allow coverage of the whole earth. The bulk of the satellite data comes from the first two groups, with very few satellites in polar orbit.

Chapter 13. The Way We Are And Where We Are Heading

A great deal of data is collected by satellites studying the earth, and this is called remote sensing. This data has a great many uses, but it also has limitations.

In the late 1960s, NASA had a large geological group studying the earth and moon. NASA also had a large budget for preparations to go to the moon. To justify the rapid money burn rate, NASA geologists tried to persuade petroleum geological societies that remote sensing would soon be doing our jobs. In one of those presentations, the NASA geologist said that we might as well all quit, because NASA was going to find all the oil with remote sensing. As it turned out, in petroleum geology, remote sensing was never more than a fair reconnaissance tool. It never found a barrel of oil.

Before speculating about how we will live in the future, it is important to discuss the accuracy of future projections. Many published climate projections are based on computer-generated climate simulation models. We have all seen the amazing things computers do when using factors that are accurately quantified. They allow us to design complex airliners and put laboratories in space.

However, computers cannot effectively handle factors that cannot be quantified. Essentially no factors involved in climate models of the future can be quantified. The sun's energy output is variable and not predictable. Concentrations of water, water vapor, and ice in the atmosphere are constantly changing; they are very important to the climate, and are unpredictable. Oceans are a huge storehouse of energy and are very complex, important to the climate, and not well quantified. What goes on at the interface between atmosphere and oceans, and atmosphere and land, are not very well understood and are not quantifiable.

Local weather forecasts use computer methods similar to climate models, but are not nearly as complex. In the central part of the country, the eighth day of an eight day weather forecast has about a 65% chance of being correct. One-hundred-year climate model forecasts based on computer programs are like the 36,500th day of a local forecast and are basically a wild guess to four decimal points.

Long term climate models are not predictive. Their value is in studying mechanisms in the system. Computer models are hypnotic and when people work on them for a long time, they begin to believe them. Some people enjoy reporting sensational stuff, so they jump on climate models like a robin on a worm.

Sometimes People Can be an Arrogant Lot

There are times when some humans believe that we are in control. They think we cause most things that are good or bad. If there is a hot summer, then humans surely must be the cause. If four or five hurricanes hit the Gulf of Mexico in one year, then we did something to cause it. The media often perpetuates this thinking, because it is cheap and easy filler and increases the audience.

It's magic that the effects of low carbon dioxide have not been recognized.

When a magician performs a magic trick, our attention is diverted from what is really going on by the magician's extraneous activity. We follow what he shows us with one hand while the other hand is making the real change. In my opinion, this is what happened with the effects of low atmospheric carbon dioxide near the end of the last ice age. We have been so preoccupied with the possible effects of the rise in atmospheric carbon dioxide that we have ignored the devastating effects of low atmospheric carbon dioxide near the end of the last ice age.

There is an increase in atmospheric carbon dioxide. There is a greenhouse effect. We may currently have minor global warming. These are facts that may or may not be related. If you ignore geological history and use your imagination, you can believe a theory that says: with an increase in atmospheric carbon dioxide, you get an increase in the greenhouse effect, which produces global warming.

Never mind that you are dealing with minute changes in the composition of the atmosphere, with only a rise of 80 parts per million carbon dioxide in the last 150 years, from 280 PPM to 360 PPM. Never mind that the carbon dioxide content of the atmosphere is very close to being the lowest in geological history, with the lowest occurring about 14,000 years ago. Never mind that water in the atmosphere, ranging from 10,000 PPM to over 40,000 PPM, is a far larger part of the greenhouse effect than 360 PPM carbon dioxide. Also, the changes of water content in the atmosphere are very significant and rapid, but we have no way to measure them.

Never mind that the apparent increase in global temperature is probably caused by an increase in the heat island effect over the last hundred years in the cities, towns, and airports worldwide where the temperatures are measured. Never mind that any changes caused by the greenhouse effect are imprinted on dramatic changes associated with the glacial cycle. Never mind that minor changes in the energy output of the sun are far more significant than anything that happens in the atmosphere.

It seems to me that the overexposure in the media of the theory of global warming from the increase in greenhouse gases has masked the real long-term problem. That problem is how low the atmospheric carbon dioxide is in relation to the needs of plants, which are our food source.

Chapter 14. The Future

All our recent evaluation deals with increasing atmospheric carbon dioxide, population, and wealth. The future will be dealing with deteriorating land use, decreasing atmospheric carbon dioxide, and decreasing population of plants, animals, and humans.

We live in a dynamic and ever-changing world and humans do many things that make the future more complex and unpredictable. However, there are some basic events we know will happen, but the timing is uncertain. An educated guess will be provided on the timing, but guess is the keyword.

Fossil fuel usage will probably peak in about fifty years with oil usage peaking earlier. Now we'll explore life with declining fossil fuel usage, and then life without fossil fuels (which may come in about 300 years). Human population should peak in about fifty years. A declining and aging population will provide challenges that will be looked at. Atmospheric carbon dioxide should peak in 50 to 100 years and should decline rapidly within 300 years. Life with declining atmospheric carbon dioxide will be explored. There will also be speculation on life in the coming glacial cycle.

Life in the Age of Declining Fossil Fuels

The fossil fuels consist of oil, natural gas, and coal with tar sands and oil shale contributing to oil. Currently in the United States fossil fuels provide about 85% of energy consumed and the world percentage is about the same. The reserve life of each fossil fuel is very different. Reserve life is the number of years' supply of proved reserves if consumption continued at the same rate as today. Of course, production and consumption continues to increase, which shortens the actual production life. Proved reserves do not include fossil fuels that will be found in the future. Those reserves yet to be found will increase the production life. Reserve life of conventional oil is now about 30 years; for natural gas about 60 years; and coal about 250 years.

With the production peak of total energy from all three fossil fuels occurring in about fifty years, there will be a slow decline from that point on. During this decline, most of the production will be natural gas and coal, as oil will be depleted first. All fossil fuels will be depleted in about 300 years.

The decline in fossil fuels will cause shortages and the price of fuel will go up. When the price goes up, the poorer countries cannot compete and will be the first to feel the shortages. They may return to old farming methods that do not use fossil fuels. The food distribution system would suffer without fossil fuels and major cities would likely be short of food and other goods.

In the richer countries, the governments will presumably take action to spread the costs and mitigate some of the problems. History may give some insight into what might happen. In the United States during the Second World War, gasoline was rationed and a national speed limit was set at thirty-five miles per hour. In addition, there were essentially no new tires for civilians. Only retread tires were available. During the shortages in the late 1970s, the national speed limit was set at fifty-five miles per hour. This increased the energy efficiency of the vehicles. It also greatly reduced the productivity of the transportation industry (while it did major damage to the leisure and travel industries).

With declining fossil fuel supplies, many industries that depend on fuel, such as airlines, will be in deep trouble and will contract. The industrialized nations that use lots of fossil fuel will have to become more efficient and probably reduce usage. Many products will have to be manufactured closer to the consumer to cut down on shipping costs. The design of most factories and other facilities will need to be changed to use passive solar energy. In the first half of the twentieth century, most of the factories had sky lights to capture the most heat and light; design features like that should see a comeback.

Many electrical utilities will have to change the way they do business. Some oil-fired plants will have to be converted to use coal. Some of the natural gas utilities will also have to convert to coal. Nuclear fission plants will generate a higher percentage of the electricity. With a little bit of luck, nuclear fusion plants will be up and running by the time fossil fuels usage is declining.

In fifty years, vehicles may be running on oil coming from oil shale, tar sands, or a liquid fuel derived from coal. If nuclear fusion is a large and economical producer of electricity, then hydrogen from water produced by electrolysis may be a major vehicle fuel. In some areas, bicycles may make a comeback as a means of transportation.

Migration from the cold northern cities to the warmer southern areas should increase as the cost of fuel increases or becomes unavailable. This will disrupt the economics of the northern cities and put a major burden on the southern areas.

China and India will be the biggest consumers of fossil fuels with the United States and Europe being third and fourth. The problems they will face with their large populations will be much greater than what we will face.

Farming in North America and Europe uses more energy than any other part of the world. Consequently, the declining fossil fuels will have the greatest impact on these areas. Today about half the world uses farming methods that require little or no fossil fuels. There will be little impact on these areas.

In about fifty years, the world's population will peak and start to decline. This will soften the effect of declining fossil fuel.

Life in a Time of Declining and Aging Population

There are many projections on when the world population will peak and at what level. Most people think the population will peak in the year 2050. The estimates 15 years ago for the peak population ranged from 9 to 12,000,000,000 people. In 2006, the estimate for peak population is 9,000,000,000 or a little less. The estimate has been dropping almost every year. The biggest factors in these falling estimates are disease and birth control in the lesser developed nations. Famine and quiet wars, ignored by the press, are significant factors. Increased production of ethanol from corn in the United States reduces food exports to many countries.

The population trends in Japan and parts of Europe give a preview of what will happen worldwide when the population declines after 2050. Germany, Lithuania, and Ukraine have the lowest birthrates. Japan has had an aging, decreasing population for some time. The population of Japan is expected to fall over 20% between now and 2050. In Germany, the population is expected to fall about 10% between now and 2050.

The aging and declining population has a major effect on social services. When there is an older population with more people retired, there are fewer workers to support retirement benefits. With fewer people paying taxes, the government has no option but to cut social services and benefits. Most countries are not prepared to deal with this situation, and in most cases private investment will not fill the gap.

The aging population will require more health care and personal care. A larger part of the workforce will have to be devoted to taking care of the old folks. Each country will have to decide how much of the workforce to use to take care of the elderly. Health care is very expensive and the cost of health care is going up faster than most other things. This will be a major part of many national budgets.

Today, there is not much discussion of euthanasia. In a time of declining and aging population coupled with declining fossil fuels and other resources, euthanasia will probably play a bigger role. How it is used and carried out will be decided when the time comes.

All of this is not just an academic exercise that has little or no meaning. These problems will affect most people under the age of thirty today and some people as old as fifty. Unlike euthanasia, the solution to many of these problems cannot wait until the time comes. They need to be addressed as soon as possible. Long-term funding of social security needs to be put in place now. Medicare needs to be viable and sustainable for many decades.

The population decline starting in the year 2050 will not be uniform and the aging of the populations will be very different in each country. Between now and 2050, the greatest population growth will be in those countries closer to the equator. This would suggest that the decline in population, after the peak is reached, would also be in the countries closer to the equator. These countries will have a very young population between now and 2050. The average age of these populations should rise when population decreases.

The countries in the mid and high latitudes are much more dependent on fossil fuels. After the peak of fossil fuel use is past, the population of the mid latitude countries will diminish. The average age of these people will rise as the population decreases. Life will get tougher in the mid and high latitudes as we run out of fossil fuel.

Life After the End of the Fossil Fuels

Fossil fuels are a depleting resource and will be essentially gone in about 300 years. A large number of factors could change the number of years left, but the outcome is certain. We will run out of fossil fuels. In looking at history, all great civilizations have ended — some with a bang and some with a whimper. Some of them transition into the next phase without major disruption. The fossil fuel age is different from any other age because it is truly global. The age of fossil fuels will have lasted between 450 to 500 years.

The end of fossil fuels will come with a very large price increase as the supply runs out. There will probably be minor production from deposits that are com-

Chapter 14. The Future

mercially viable only at very high prices to make lubricants and petrochemicals and other high value products.

It is very difficult to visualize life without fossil fuels. They are a major part of all facets of life. The best way to view the end of fossil fuels is that it is the reverse of the start. Before the start of the fossil fuel age, energy came from other sources. Muscle power, both human and animal, was a major source of energy. Wind was the most important source of power for moving ships and cargo on water. On land, wind pumped water from wells and provided energy to grind grain. The energy from moving water in streams and rivers was captured by water wheels which provided the energy for early manufacturing. The first steam engines burned wood for the power source. Home heating was provided by wood burning stoves. The sun heated homes by what is now called passive solar heat.

About 150 years ago, fossil fuels started to replace other forms of energy. Coal in America and peat in England were the first fossil fuels to replace wood in many applications. Oil products were used for lighting in kerosene lamps and home heating. Natural gas was also used in gas lights and heating. Next, fossil fuels replaced wood to fuel steam engines used in transportation.

After fossil fuels replaced existing power sources, new uses for fossil fuels were created. Electricity was developed, and fossil fuels were the energy source used. Individual vehicles and farm machinery were developed to run on oil products. As the gasoline became better, air travel became possible. Along the way, petrochemicals and coal tar product became a vital, but little recognized, part of our life.

Within 300 years, all fossil fuels have to be replaced or the functions they enable will come to a stop. A major use of coal, oil, and gas is to generate electricity. Their biggest competitor is nuclear power plants which are fueled by uranium, a depleting resource. One solution is to use breeder reactors which produce more fuel than they consume. Nuclear fusion may be commercial by then. Fossil fuel power plants will probably be the first use to be replaced.

Fuel for airplanes will be one of the most difficult to replace. They will probably require entirely new kinds of engines. New fuels and engines will be needed for individual vehicles and farm machinery. New ways to heat homes will have to be developed, and passive solar may be a big part of that.

The good news is that people have close to 300 years to make the transition from fossil fuels to the next fuels, whatever they may be. If adequate fuels do not become available, then that 300 year lead time should be used to reduce reproduction so that the human population is adjusted down to what the world can feed. Easy to say.

There will probably be huge changes in the way businesses function without, or with severely reduced, fossil fuels. People will have to shorten their commute or work at home. Supply lines will probably shorten and products will be made closer to the consumer.

The way humans are, there will probably still be wars. However, the way wars are fought wars will have to change after the fossil fuels run out.

When we no longer have fossil fuels to burn, we will stop putting carbon dioxide into the atmosphere. When this happens, more of the atmospheric carbon dioxide will be absorbed by the oceans and the atmospheric concentrations of carbon dioxide will decrease.

Life in the Age of Declining Atmospheric Carbon Dioxide

To explore life when there is declining atmospheric carbon dioxide, we will make some assumptions that will change when better data is available. Atmospheric carbon dioxide should peak in 50 to 100 years and then start to decline. The peak atmospheric carbon dioxide should be about 200 parts per million greater than the current concentrations.

When all the accessible fossil fuels have been burned, the total increase of carbon dioxide in the ocean-atmospheric system should be in the 3% to 5% range. When everything is in balance, there is an equilibrium in the ocean–atmospheric system, where carbon dioxide given off from the warm waters near the equator balances with the carbon dioxide absorbed by the cold waters near the Poles.

With the burning of fossil fuels, carbon dioxide has been added to the atmosphere faster than it can be absorbed by the oceans. Consequently, the ocean–atmospheric system, with respect to carbon dioxide, is way out of balance. It will probably be out of balance until after all the fossil fuels are burned in about 300 years. Near equilibrium should be reached 100 years after that, as the excess carbon dioxide in the atmosphere is absorbed by the oceans. This would be 400 years from now.

At equilibrium, the atmospheric carbon dioxide level would be about 300 parts per million. This would be about 5% greater than the level during most of the interglacial period and about equal to the level in 1900. At this point, the level of atmospheric carbon dioxide would once again be dependent on the temperature of the surface of the oceans.

With this level of atmospheric carbon dioxide, vegetation in the high latitudes would diminish and less land would be capable of producing crops. Some marginal land regardless of latitude would stop producing. History suggests that the earth can support about 2,000,000,000 people when the atmospheric carbon dioxide level is about 300 parts per million.

Food production will start to be affected by declining atmospheric carbon dioxide near the end of the fossil fuel age. The problems will be double for the farmers of this time as they will have to deal with declining productivity from less carbon dioxide and loss or changing fuel for farm machinery. This will require completely new technology and a short time of transition.

Reduced food production from declining atmospheric carbon dioxide will affect some parts of the world more than others. The countries near the equator will be the least affected because more atmosphere and carbon dioxide are found there. The effect of lower atmospheric carbon dioxide will be a reduction of yields. There may also have to be a change in some of the crops planted.

In the mid latitudes, there will be a major reduction in crop yields from less atmospheric carbon dioxide. There will be a major change in farming methods from disappearing fossil fuels. There will be a change in the type of crops planted in many places. New crop varieties will have to be developed to get by with less atmospheric carbon dioxide. The farms will not produce as much as they did in the past. Many of the farms in the high latitudes will become non-productive and have to be abandoned. Those that do survive will produce far less.

Patterns of food consumption and distribution will change. The people that live near the equator will need to survive on what they produce. There will be no surplus food grown in the mid latitudes that can be sent to the countries near the equator. Some food grown near the equator may be exported to the mid latitudes.

In the mid latitudes, there probably will be food shortages. If this occurs, the urban poor will be the losers. The rural areas will feed themselves before they ship anything to the cities.

LIFE IN THE COMING GLACIAL CYCLE

In the last 2,000,000 years, there has been a glacial cycle about every 100,000 years. Of these, four have been very large as determined by field evidence of the glacial deposits. The last glacial cycle was very large, and this would suggest that the next one may be milder.

The climate change will be caused by changes in the energy output of the sun and astronomical variations in the energy delivered to earth. There will be continuing energy buildup in the oceans which is necessary to evaporate the top several hundred feet of the oceans to build the glaciers when the accumulation phase comes. Today we can predict the astronomical variations affecting the energy input, but we cannot predict the variable energy output of the sun

About 10% of the earth's land surface is covered by ice today. In the last glacial cycle, about 27% of the land surface was covered by glaciers, which gives us a rough estimate of the maximum for the next glacial cycle. As the glaciers expand in the next accumulation phase, the climate will change in many areas of the earth. Some deserts in the lower latitudes will become cooler and receive more rain, which will be positive. Most areas, however, will be colder and will receive more rain and snow. These changes will be negative for most plants, animals, and humans. With less land and a colder, harsher climate, the population of plants, animals, and humans will decrease significantly. The environmental conditions will deteriorate for about 50,000 years throughout the accumulation phase and well into the dormant phase.

The environmental conditions for plants, animals, and humans will start improving late in the dormant phase. In the active phase, the climatic conditions will improve on the land areas not covered by glaciers. There will be an increase in precipitation and atmospheric carbon dioxide from the very low levels of the dormant phase. With these improving conditions, the population of plants, animals, and humans will increase. This situation is relatively short-lived and will change in the next glacial breakup phase.

The next glacial breakup phase will be very hard on living things because of a major decrease in atmospheric carbon dioxide and lower precipitation. However, I suspect that the next breakup phase will not cause the extinction of many animal genera like the last one did, for several reasons. With all of the fossil fuels burned, there will be an increase of carbon dioxide in the ocean-atmosphere system. Many compensating mechanisms that will affect the carbon dioxide in the system include deposition of carbonates in shallow seas and utilization by ocean plants. In addition, we do not know how much carbon dioxide will be added to the system by volcanism.

Glacial cycles after the next one are too remote for us to make any viable predictions. We do not know if there will be another great extinction with the same cause as the one that occurred near the end of the last ice age, but it is a possibility.

Conclusion

There have been major and continuing climate changes throughout the Pleistocene during the last 2,000,000 years. These changes came from the many glacial cycles and affected all plants, animals, and humans. Energy from the sun is the power source of essentially all of the earth's mechanisms and of life. Changes in the amount of energy reaching the earth from the sun are the root cause of the many changes during the glacial cycle, including changes in plants, animals, and humans.

Vast amounts of energy stored in the oceans are required for large continental glaciers to form and accumulate. Continental glaciation breakup occurs only when there is a very large amount of energy input to the earth's atmosphere that melts the glaciers. The atmosphere, oceans, land, and glaciers are all interconnected. Changes in any one of these affect the others, and all life.

All life is dependent upon the survival and viability of plants. Plants on land need sufficient atmospheric carbon dioxide to survive. Carbon dioxide is very soluble in water, and is about twice as soluble in cold water as it is in warm water. The carbon dioxide in the atmosphere varies with the temperature of the surface of the oceans. When the oceans' surfaces were cold near the end of the last ice age, the atmospheric carbon dioxide dropped dramatically. Plants could not survive in the higher altitudes, middle and high latitudes, and some plants could not survive in the low latitudes. The animals and humans that ate these plants also did not survive.

In this book, we have looked at net effects of these phenomena. The percentage of carbon dioxide in the atmosphere has declined throughout most of geological time. The atmosphere is deformed by centrifugal force. Most of the energy reaching the earth is utilized in changing the temperature or phase of water. Plants cannot grow when they do not have sufficient carbon dioxide, water, nutrients, and sunlight. Animals cannot survive without sufficient plants for food. All of these mechanisms are much more complex than we can identify, but we can see the net effects.

In the last 20,000 years, there have been three periods of time when the atmospheric carbon dioxide levels have changed rapidly. Each of these changes had major effects on the life and distribution of plants and animals. The well-being of humans was altered dramatically by each change in the level of carbon dioxide in the atmosphere.

The extreme drop in atmospheric carbon dioxide near the end of the last ice age greatly reduced the land area where plants could grow. Over half of the large animal genera became extinct. Humans could no longer survive in many parts of the world, and the human population was greatly reduced where they did survive.

After the last ice age was over, and the surface temperature of the oceans rose, there was a large and rapid increase in atmospheric carbon dioxide. Plants and animals multiplied and migrated to the mid and high latitude areas of the world. There was a population explosion of humans and they moved with the plants and animals to the mid and high latitudes.

Today, we are experiencing the fastest ever rise in atmospheric carbon dioxide, accelerated by the burning of fossil fuels. Food production from domestic plants has increased at a very rapid rate. There has been a large increase in wild plants and animals where humans have not restricted the increase. The human population has increased over sixfold in the last 150 years, and it is still continuing to increase. The rapid increase in atmospheric carbon dioxide that is occurring now and powering all these developments will be followed by a rapid decrease.

The large animal extinction near the end of the last ice age demonstrates that the earth is not always a safe and secure place. We humans living in highly developed areas of the world have come to believe that being healthy, safe, and secure is a God-given right. We seem to think that if things do not go well, then the government or someone else is to blame. However, life on earth comes with no guarantees. We are responsible for our own well-being. We are also partly responsible for the well-being of the wild plants and animals on earth. However, we must realize that in most respects we are not in control. We are just passengers on a multifaceted and changing earth, which is far more complex than humans will ever fully understand.

Sources

Agenbroad, L.D. and J.L. Mead, Editors. 1994. *The Hot Springs Mammoth Site.* Hot Springs, Dakota: The Mammoth Site of South Dakota.

Allen, J.E. and M. Burns. 1986. *Cataclysms on the Columbia.* Portland, Oregon: Timber Press.

"Ancient Grain Varieties in Archaeology." 2005. *Newarchaeology.* 9 September 2006. ‹http://www.newarchaeology.com/articles/ancient_grain.php›.

"Ancient Irrigation." Geology University of California at Davis. 2 September 2006. ‹http://www.geology.ucdavis.edu/~cowen/~gel115/115CH17oldirrigation.html›.

Ballard, R.D. 1983. *Exploring our Living Planet.* Washington, D.C.: National Geographic Society.

"British Empire." *Columbia Encyclopedia,* sixth edition. 2001-2005. 29 September 2006. ‹http://www.bartleby.com/65/br/britemp.html›.

Buchanan, R. Editor. 1984. *Kansas Geology: An Introduction to Landscapes, Rocks, Minerals, and Fossils.* Lawrence, Kansas: University Press of Kansas.

Burenhult, G. Editor. 1993. *The First Humans: Human Origins and History to 10,000 Years BC.* New York: Harper Collins.

"Carbon Dioxide." *Wikipedia, the Free Encyclopedia.* 21 August 2006. ‹http://en.wikipedia.org/wiki/Carbon_dioxide›.

Carnegie Library of Pittsburg. 1994. *The Handy Science Answer Book.* Detroit, Michigan: Visible Ink Press.

Denton, G.H. and T.J. Hughes, Editors. 1981. *The Last of the Great Ice Sheets*. New York: John Wiley and Sons.

Eagleman, J.R. 1985. *Meteorology: The Atmosphere in Action*. Belmont, California: Wadsworth.

Friedman, G.M. and J.E. Sanders. 1978. *Principles of Sedimentology*. New York: John Wiley and Sons.

Gilbert, C.M. and R.L. Brooks. 2000. *From Mounds to Mammoths: A Field Guide to Oklahoma Prehistory*. Norman, Oklahoma: University of Oklahoma Press.

Hammond Atlas of the World. 2003. New York: Barnes and Noble Publishing.

Heacox, K. 1992. *Alaska's Inside Passage*. Santa Barbara, California: Albion Publishing Group.

Herzog, D.C. 1992. "Hazards of Geomagnetic Storms." *Earthquakes and Volcanoes* 23:#4 152-159.

Hodgeman, C.D. Editor. 1951. *Handbook of Chemistry and Physics*. Cleveland, Ohio: Chemical Rubber Publishing.

"How Much Natural Gas is There?" 2006. *NaturalGas.org* 6 November 2006. ⟨http://www.naturalgas.org/overview/resources.asp⟩.

Imbrie, J. and K.P. Imbrie. 1979. *Ice Ages Solving the Mystery*. Short Hills, New Jersey: Enslow.

Judson, S. and M.E. Kauffman. 1990. *Physical Geology*. Englewood Cliffs, New Jersey: Prentice Hall.

Kansas Association of Wheatgrowers. "Kansas Wheat Sketches 1918-1998." 2 September 2006. ⟨http://wheatmania.com/Grainsofhistory/wheatsketches.htm⟩.

Keller, W.D. 1957. *The Principles of Chemical Weathering*. Columbia, Missouri: Lucas Brothers.

Kucera, R.E. 1978. *Probing the Athabasca Glacier*. Vancouver, British Columbia: University of British Columbia.

Kuenen, P.H. 1950. *Marine Geology*. New York: John Wiley and Sons.

Lethcoe, N.R. 1987. *Glaciers of Prince William Sound, Alaska*. Valdez, Alaska: Prince William Sound Books.

"Little Ice Age." *Wikipedia, the Free Encyclopedia*. 13 September 2006. ⟨http://en.wikipedia.org/wiki/Little_ice_age⟩.

Lyons, W.A. 1977. *The Handy Weather Answer Book*. Detroit, Michigan: Visible Ink Press.

Sources

MacDonald, G.A. and D.H. Hubbard. 1982. *Volcanoes of the National Parks in Hawaii*. Hawaii Volcanoes National Park, Hawaii: Hawaii Natural History Association in Cooperation with the National Park Service.

Mandia, S.A. "The Little Ice Age in Europe." 21 September 2006. <http://www2.sunysuffolk.edu/Mandias/lia/little_ice_age.html>.

Marler, G.D. 1964. *Studies of Geysers and Hot Springs Along Fire Hole River, Yellowstone National Park, Wyoming*. Yellowstone Library and Museum Association.

Martin, P.S. and R.G. Klein. Editors. 1985. *Quaternary Extinctions: A Prehistoric Revolution*. Tucson, Arizona: The University of Arizona Press.

Martin, P.S. and H.E. Wright, Jr. Editors. 1967. *Pleistocene Extinctions: Search for a Cause*. New Haven, Connecticut: Yale University Press.

McGeveran, W.A. Editor. 2006. *The World Almanac and Book of Facts 2004*. New York: World Almanac Books.

"Methane Clathrate." *Wikipedia, the Free Encyclopedia*. 25 August 2006. <http://en.wikipedia.org/wiki/Methane_hydrate>.

Moore, R.C. 1933. *Historical Geology*. New York: McGraw-Hill.

Moore, R.C. 1949. *Introduction to Historical Geology*. New York: McGraw-Hill.

Moore, R.C., C.G. Lalicker, and A.G. Fisher. 1952. *Invertebrate Fossils*. New York: McGraw-Hill.

Morin, B. 1984. "The Greenhouse Effect." *Nature Canada* 13 (October/November).

National Park Service Handbook. 1983. *Glacier Bay: A Guide to Glacier Bay National Park and Preserve, Alaska*. Washington, D.C.: Division of Publications National Park Service.

Nilsson, T. 1983. *The Pleistocene: Geology and Life in the Quaternary Ice Age*. Dordecht, Holland: D. Reidel.

"Oil Shale." *American Association of Petroleum Geologists*. May 2005. 6 November 2006. <http://emd.aapg.org/technical_areas/oil_shale.cfm>.

Oxford Atlas of the World. 1993. New York: Oxford University Press.

Post A. and E.R. LaChapelle. 2000. *Glacier Ice*. Seattle, Washington: University of Washington Press.

Rennick, P. Editor. 1993. "Alaska's Glaciers." *Alaska Geographic*. 9 #1 Revised.

Rennick, P. Editor. 1994. "Prehistoric Alaska." *Alaska Geographic*. 21 #4.

Snook, J.K. 1996. *Mental Wanderings: The Book for People with Short Attention Spans*. Salt Lake City, Utah: Northwest.

Thornbury, W.D. 1958. *Principles of Geomorphology*. New York: John Wiley and Sons.

Ukraintseva, V.V. 1993. *Vegetation, Cover, and Environment of the "Mammoth Epoch" in Siberia*. Hot Springs, South Dakota: The Mammoth Site of South Dakota.

United States Department of Energy. "Oil Shale Activities." 6 November 2006. ‹http://www.fe.doe.gov›.

United States Geological Survey. "Geology of the Loess Hills, Iowa." 9 August 2006. ‹http://pubs.usgs.gov/info/loess›.

"Water Resources." *Wikipedia, the Free Encyclopedia*. 26 October 2006. ‹http://en.wikipedia.org/wiki/water_resources›.

Weiner, J. 1986. *Planet Earth*. New York: Bantam.

Yokobori, K. 2006. "Survey of Energy Resources." *World Energy Council*. 6 November 2006. ‹http://www.worldenergy.org/wec-geis/publications/reports/ser/overview.asp›.

Index of Names

A

Agassiz, Louis, 5
Alaska Geographic, 178
Alaskan Peninsula, 13
Aleutian Islands, 51
Amazon River, 25, 54
American Association of Petroleum Geologists, 177
Ancient Irrigation, 115, 138, 175
Andes, 19, 25, 86
Antarctic Oceans, 51
Antigua, 122
Aquifers, 158, 161
Archaeology, 175
Arctic National Wildlife Refuge, 147
Arctic Ocean, 50, 53, 56
Arkansas River, 136
Artifacts, 98, 101, 107, 109-110
Asian Russia, 13
Aswan Dam, 138
Athabasca Glacier, 23, 176
Athabasca Tar Sands, 148
Atlantic Ocean, 50, 55, 110
Atoka Formation, 152
Axel Heiberg, 120

B

Baffin Islands, 120
Bahama Banks, 12
Baltic Sea, 12, 74
Baranof Island, 88
Barents Sea, 13
Bay of Bengal, 56, 112
Bering Strait, 56
Black Sea, 12, 50, 56
Bravo Dome, 60
British Empire, 175

C

Cambrian, 91-92
Canadian Arctic Islands, 13, 120
Canadian Rockies, 72
Caribbean Sea, 12, 56, 109, 112
Carlsbad Caverns, 52
Celtic Sea, 12
Chile, 5, 13, 25, 120
Chukchi Sea, 13
Clayton NM, 60
Clovis, 110
Coelacanths, 93
Cold Lake, 149
Colorado Plateau, 152
Colorado River, 158

Columbia Glacier, 119, 176
Columbia River, 79
Continental Divide, 13, 136
Cordilleran, 55, 106
Current Interglacial Period, xiii, 5, 112

D

Denali National Park, 51
Dirty Thirties, 86

E

East Antarctica, 6, 9, 22, 54-55, 87, 99, 108
East China Sea, 12
Egypt, 117, 138
El Niño, 50, 162
Ellesmere, 120
Euphrates, 116-118

F

Falkland Islands, 12
Flint Hills of Kansas, 137
Full Data Diagram, ix, xii, 44-45
Fusselman Formation, 150-151
Fusselman Gas Well, 151

G

Galapagos Islands, 93
Geysers Field, 154
Glacier Bay, 119, 136, 177
Gondwanaland, 19-20
Grand Banks of Newfoundland, 12
Grand Coulee, 74
Grazers, 18, 113
Great Canadian Oil Sands Company, 148
Great Plains, 76, 86, 94, 144, 152
Great Sand Dunes National Park, 77-78
Gulf of Mexico, 12, 55-56, 109, 112, 136, 164

H

Halemaumau Lava Lake, 60-61
Hawaii Volcanoes National Park, 177
Himalayas, 117
Hot Springs Mammoth Site, 175

Hudson Bay, 13, 72, 74, 122-123
Hudson Bay Company, 122-123
Hugoton Gas Field, 127, 132
Hwang Ho River Valley, 117

I

Iceland, 13, 36, 53, 120, 122, 154
Indian Ocean, 56, 93, 99
Indonesia, 12, 106, 108, 112, 121
Indus River Valley, 117
Inside Passage, 176
Inuit, 109

J

Java, 69, 104

K

Kansas Agriculture Statistics Service, 131
Kara Sea, 13
Kerogen, 149
Kilauea Volcano, 60-61, 156
Kilimanjaro, 39, 86
Kluane National Park Reserve, 120
Krakatoa, 69-70

L

LaBrea Tar Pits, 24
Lake Bonneville, 73-74
Lake Nasser, 138

M

Madagascar, 102
Maidenhair, 83
Mammoth Site of South Dakota, 175, 178
Marine Geology, 176
Mastodon, 3, 19, 23
Mauna Loa Volcano, 156, 161
Maunder Minimum, 33, 35, 39, 121
Mears Glacier, 119
Medieval Warm Period, 118
Mediterranean Sea, 12, 50, 56, 110, 116, 118
Melt Water Distribution, xii
Mesopotamia, 116-117

Methane, xiv, 60, 151-152, 177
Methane Clathrates, xiv, 151-152
Mezhirich, 18
Mid Atlantic Ridge, 36
Middle Eastern Oil Reserves, 152
Mississippi Delta, 79, 136, 138
Mississippi River Valley, 79
Mississippian, 82
Mitochondrial DNA, 109
Moore-Hooper, 60, 151
Moore-Hooper Ellenburger, 60
Moraine Lake, 72
Mount Kilimanjaro, 39

N

NASA, 163
National Park Service, 177
Natural Gas, xiv, 52, 60, 83, 127, 132, 145-146, 148-151, 165-166, 169, 176
Neanderthals, 105-107
New Guinea, 12-13, 16, 106-107, 112
New Orleans, 78-79, 85
Nile Delta, 117
Nile River Valley, 117
Nitrogen, 8, 36, 50, 60-63, 100, 150
North Atlantic Ocean, 55
North Pacific Ocean, 56
North Sea, 12-13, 66, 74
Nuclear, xiv, 36, 146-147, 153, 166, 169

O

Ogallala Formation, 131
Oil Shale, xiv, 149, 165-166, 177-178
OPEC, 147
Origin of Loess, xii, 76
Orinoco, 148
Outer Continental Shelf, 147

P

Paleozoic Era, 82
Peace River, 149
Periglacial, 13, 39, 84
Philippine Sea, 56
Pleistocene Extinctions, 177
Precambrian, 91
Prehistoric Alaska, 178

Prince William Sound, 119, 176
Prudhoe Bay, 139

Q

Quaternary Extinctions, 14, 16, 97, 177
Quaternary Ice Age, 177

R

Rocky Mountains, 5, 136, 140
Russian Arctic, 13, 120
Russian Arctic Islands, 120

S

Sahara, 116
Saint Elias Mountains, 120
Saint Lawrence Seaway, 74
San Juan Mountains, 77
San Louis Valley, 77
Sand Storm Effects, 29
Sangre de Cristo Mountains, 77
Seashores Sinking, xiii, 136
Siberia, 11, 13, 18, 107, 178
Silurian, 82
South Africa, 19, 26, 39, 56, 153
South Asia, 6, 19, 26, 88, 94, 97
South Atlantic Ocean, 55
South China Sea, 12, 56, 112
South Pacific Ocean, 55
Strait of Gibraltar, 12
Studies of Geysers, 177
Sundra Strait, 69
Sunspots, 33, 121
Swiss Alps, 121
Switzerland, 5, 122

T

Tambora Volcano, 121
Tar Sands, xiv, 83, 148-149, 165-166
Tasmania, 12-13
Thames River, 121
Tidewater, 38-39, 43, 119
Tigris River, 116
Tongass National Forest, 88
Tree Lines, xii, 51, 65-67, 76, 85-86, 96, 100
Triassic, 93

Twilight, 70
Types of Glaciers, xii, 38

U

Unconformities, 111
Uranium, 34, 109, 153, 169

V

Valdez, 119, 139, 176
Valley of Ten Peaks, 72
Venezuela, 19, 25, 110, 148
Venice, 136

W

Wastage, 38, 41, 46, 53, 67

Water Resources, 132, 178
Way We Are, xiii, 143
West Texas, 22, 132, 150-151
Western Canada, 13
Western Europe, 114, 132
Western North American, 23
Wind, xiv, 7, 19, 27, 29, 35, 65, 69-70, 76-77, 81, 86, 97, 130, 134, 144, 146, 155, 169
Wood, xiv, 106, 108, 114, 146, 154, 161, 169
Woolly Mammoth, xi, 3, 17-18, 23, 25, 95

Y

Yellow Sea, 12
Yellowstone National Park, 36, 69, 177
Yosemite Falls, 72
Yukon, 15, 120

Index of Subjects

A

Absolute Humidity, 179, 40
Accuracy of future projections, 163
Aerial photography, 120
African lion, 24, 26
African long-horned buffalo, 26
African survivors included giraffe, 26
Absolute Humidity, 40
Accumulation Phase, xii, 33, 41, 43-44, 50, 78, 83-84, 94, 143, 171-172
Active Phase, xii-xiii, 41, 46, 53-54, 67, 73, 78, 84-85, 95, 99, 107, 172
Age of Declining Atmospheric Carbon Dioxide, xiv, 170
Age of Declining Fossil Fuels, xiv, 165
Age of fossil fuels, 146, 168
Aging Population, 165, 167-168
Agriculture acreage, 157
Air density, 40, 66, 100
Air inclusions, 66, 100, 111
Air packs, 150
Alaska's Glaciers, 178
Alaska's Inside Passage, 176
Alaskan Peninsula, 179, 13
Albedo chart, 49
Albedo of the glaciers, 77
Albedo of Various Surfaces, 37
Amazon water, 55
American mastodons, 19

American Pleistocene lion, 24
Amount of energy, 7, 33-34, 36-38, 43, 49, 73, 173
Amount of food, 131
Amount of oil, 160
Analysis of air inclusions, 100
Ancient atmosphere, 62
Ancient dust storms, 29
Ancient Egyptian civilization, 117
Ancient forests, 82
Ancient Grain Varieties, 179, 175
Ancient human village, 18
Ancient ice sheets, 41
Ancient Irrigation, 179, 115, 138, 175
Andine gomphothere, 19
Animal extinction, 181, xi, xiii, 1, 4, 6, 8-9, 11, 16, 18-19, 21-24, 47, 57, 87-89, 91, 97-99, 101, 107, 113, 172, 174
Animal extinction pattern, 16
Animal genera, 3-4, 7, 20, 24, 26, 30, 94, 98-99, 101, 108-109, 113, 125, 172-173
Animal Population, xiii, 23, 102, 113, 145
Animal productivity, 126
Animal transitions, 17
Animals' food, 4, 91, 95-96, 102, 115, 144, 161
Animals' habitats, 71, 91, 97
Antarctic Oceans, 179, 51
Anthracite coal, 127
Appalachian Basin, 152
Arctic Ocean, 179, 50, 53, 56

Arctic tundra, 87
Areas of permafrost, 13
Areas of possible loess, 29
Arkansas River, 179, 136
Argon, 50, 60-61, 63
Arrogant Lot, xiv, 163
Artifact evidence, 7, 108
Artistic evidence, 106
Asian horse, 24, 26
Aspects of the climate, 104
Astronomical variations, 34, 171-172
Athabasca Glacier, 179, 23, 176
Atmospheric carbon dioxide concentration, 100
Atmospheric carbon dioxide content, 111
Atmospheric carbon dioxide effect, 181, xi, 9
Atmospheric Gases, xii, 50-51, 68, 75, 85
Atmospheric oxygen, 100
Atoka Formation, 179, 152
Atoka Formation water, 152
Aurora borealis, 35
Australian animals, 20
Australian plants, 22, 27, 87
Average ocean rise, 54
Axel Heiberg, 179, 120
Axis of rotation, 34, 63-64

B

Background Information, xii, 44
Barren land, 17, 27, 76-77, 87, 95, 98
Beach erosion, 138
Beneficial insects, 133
Bering Sea, 179, 11, 51
Bering Strait, 179, 56
Big island of Hawaii, 141
Big-tongued sloth, 24-25
Birds' eggs, 102
Birth control, 135, 145, 167
Bituminous coal mining, 127
Black Sea, 179, 12, 50, 56
Blow dirt, 86
Blue-green algae, 52, 62, 82
Bravo Dome, 179, 60
Breakup of Gondwanaland, 20
Breakup of the glaciers, 54
Breeder reactors, 153, 169
Breeding of dogs, 106
British colonies, 122

British Empire, 179, 175
Broad-faced sloth, 25
Brown bears, 24
Browsing animals, 23, 94, 96

C

California condor, 144
Calorie of heat, 36, 49, 75
Cambrian period, 91-92
Canadian Arctic Islands, 179, 13, 120
Canadian mainland, 120
Canadian Rockies, 179, 72
Canadian Yukon Territory, 120
Canals of the Netherlands, 121
Canaveral National Sea Shore, 112
Carbon dioxide absorption, 44
Carbon dioxide (amount of), 44, 52, 67, 100, 156
Carbon dioxide concentrations, 9, 51, 62, 83, 101, 170
Carbon dioxide gas, 60, 128, 150
Carbon dioxide level, 22, 43, 56, 66, 84-85, 119, 170-171, 173
Carbon dioxide removal, 51
Carbon monoxide, 60-61, 63
Carbonaceous material, 52, 152
Carboniferous, 82
Carlsbad Caverns, 180, 52
Carnivorous predators, 93
Cause of extinction, 3, 10, 172
Cave art, 106
Cenozoic, 83
Centrifugal, 9, 63-65, 100-101, 173
Chemical fertilizers, 138
Circulation patterns, 50, 79
Civilizations of Egypt, 117
Clay tills, 73
Climate models, 163
Climate projections, 163
Coal, 127
Coal tar product, 169
Colloidal sized particles, 137
Columbia Glacier, 180, 119, 176
Columbia River, 180, 79
Comet impacts, 4
Common ancestor, 104
Composition of the atmosphere, 31, 61, 164
Concentrations of carbon dioxide, 51, 62, 170

Index

Content of atmospheric gases, 68
Content of the reservoir, 150
Continental glaciers, 180, xi, 5-6, 13, 21, 32, 37-39, 49-50, 54, 56, 71-73, 75, 84, 173
Continental glaciers form, 49, 72, 173
Corn yields, 128
Creation of deltas, 79
Crop yields, 121, 144, 158-159, 171
Cuviers gomphothere, 23
Cycads, 83

D

Damming of rivers, 137
De-gassing of the planet, 60
Declining atmospheric carbon dioxide, 179, xiv, 96, 165, 170-171
Declining fossil fuels, 179, xiv, 165, 167-168
Declining population, xiv, 165, 167-168
Deep fracture systems, 60
Denali National Park, 180, 51
Depletion allowance, 148
Deposition of carbonates, 172
Depth of sunlight penetration, 75
Destruction of glaciers, 38
Development of Early Humans, 180, xiii, 103-104
Development of vegetation, 20
Devonian period, 82, 92
Diagram of Centrifugal Force, ix, 180, 64
Diminishing atmospheric carbon dioxide, 98
Dinosaur extinction, 20, 97
Disconformities, 111
Discussion of ancient irrigation, 138
Dissipating glaciers, 73, 99
Dissipation, 6, 38, 41, 71, 95
Distribution of extinction, 101
Distribution of the atmosphere, 101
Distribution patterns of vegetation, 9
Dogs, 105-106
Domestication of animals, 115
Domestication of grain, 115
Dormant phase, 41, 44-45, 51, 84-85, 95, 172
Downhole combustion, 149
Drastic reduction of plants, 109
Dust Bowl, 29, 86, 131, 141
Dust storms, 27, 29-30, 76-77, 86-88, 96-97, 108-109

E

Earliest forms of irrigation, 115
Earliest large irrigation, 116
Earth's axis of rotation, 34, 63
Earth's magnetic field, 33, 35
Earth's magnetic poles, 35
East Antarctica, 180, 6, 9, 22, 54-55, 87, 99, 108
Eastern coal, 152-153
Effect of the dust, 97
Effective irrigation systems, 117
Effects of human activity, 89
Efficient recyclers, 104
Electric power generation, 152
Emigration, 122
Energy consumption, 146
Energy of ignition, 63
Energy output, 33, 35, 39, 163-164, 171-172
Energy storage, 156
Episode of extinction, 93
Episodic events, 74
Erratics, 74
Eskers, 74
Ethanol, 134, 154-155, 167
Ethanol production, 134, 154, 167
Eurasian giant beaver, 26
Europe loess, 29
European cave lion, 26
European horse, 26
Evolution of Animals, 91
Evolution of the atmosphere, 61-62
Evolution of animals, 91
Evolution of vegetation, 41
Evolutionary processes, 93
Examples of artifact evidence, 7
Examples of parallel evolution, 93
Exploitation geologist, 140
Exploration geologist, 140
Explosion of human population, 7
External energy, 33, 36
Extinct Asian hog, 26
Extinction of animals, 4, 91, 94-95, 97-98

F

Father of glaciology, 5
Fiber production, 161

Fifteen highest dams, 138
Final instrument of extinction, 98
Final yield, 129-130
First four civilizations, 117
First Humans, 180, 67, 107, 175
Fjords of Alaska, 79
Flat-headed peccary, 24
Flint Hills of Kansas, 180, 137
Flowering plants, 83
Formation of loess, 75, 145
Formation pressure, 152
Formation water, 152
Formed by accretion, 59
Formed by blue-green algae, 82
Fossil evidence, 7, 93, 104, 107
Fossil fuel, 127, 140, 145-146, 156-157, 159-160, 165-169, 171
Fossil fuel age, 168-169, 171
Fossil fuels, end of, 168-169
Fossils, 62, 81-82, 91-93, 104-105, 109, 175, 177
Fracture systems, 52, 60
Fresh meltwater, 22, 33, 55, 100
Front of the glacier, 73
Frozen tundra, 145
Fusselman Formation, 180, 150-151
Future Projections, 180, xiv, 161, 163

G

Genera loss, 3
Genera of sloth, 25
Geographic distribution of plants, 111
Geological History, xii, 3, 82-83, 91, 100, 164
Geological time, 92, 173
Geomagnetic storms, 35, 176
Geothermal energy, 146, 154
Giant beaver, 24, 26
Giant deer, 26
Giant guinea pig, 25
Giant rhinoceros, 26
Giant short-faced kangaroo, 27
Giant wombat, 27
Glacial breakup phase, 7, 9, 39, 41, 46, 56, 66-68, 73, 75, 77-79, 85, 88-89, 96, 99, 108, 172
Glacial buildup, 22, 32, 41, 50
Glacial contraction, 6

Glacial cycle, 5-8, 10-12, 21, 27, 30-34, 41-47, 51, 55-56, 67-68, 71-72, 75, 77-79, 81-86, 89, 91, 94, 99, 104, 107-108, 143, 160, 164-165, 171-173
Glacial cycle, causes, 6, 31, 33
Glacial dome, 13
Glacial drift, 97
Glacial effects today, 71
Glacial expansion, 6, 12, 94
Glacial front, 73-74
Glacial ice cores, 66
Glacial melting, xii-xiii, 4, 6, 9, 11-12, 21-22, 55, 57, 73, 75, 87, 91, 100-101
Glacial meltwater, 33, 47, 54-55, 74, 99
Glacial movement, 41
Glacial outwash, 73, 76, 97
Glacial processes, 145
Glacial till, 73-74, 76, 114
Glaciation of the Pleistocene, 5
Glacier Bay, 180, 119, 136, 177
Glacier forms, 72
Glaciers, Continental, xi, 5-6, 13, 21, 32, 37-39, 49-50, 54, 56, 71-73, 75, 84, 173
Glaciers outside of Antarctica, 6
Glaciers today, 38-39, 119-120
Global distribution of loess, ix, 17, 29
Grain crops, 144
Grain (Ancient Varieties), 175
Grand Banks of Newfoundland, 180, 12
Grand Coulee, 180, 74
Gravity segregation, 59-60
Grazing animals, 23, 94, 96, 98, 102, 128, 157
Greatest Glacial Extent, 12
Greenhouse effect, 164, 177
Greenhouse gases, 164
Greenland colony, 122
Ground water, 32, 37, 137
Gulf Stream, 50, 162

H

Habitats of plants, 71, 144
Halemaumau Lava Lake, 180, 60-61
Hanging valley, 72
Hawaii Volcanoes National Park, 180, 177
Hazards of Geomagnetic Storms, 181, 176
Headwaters of the stream, 78
Heat Cycles, xii, 31-32
Heat of the sun, 32

Index

Heavy crude oil reservoir, 148
Herbicides, 131, 133-134, 157, 159
Herds of large animals, 93
Hermit megathere, 25
High Aswan Dam, 138
High carbon content, 83
High carbon dioxide requirement, 101
Higher carbon dioxide requirement, 87
Highest density of irrigation, 117
History of plants, 94
History of the planet, 59
Holocene vegetational pattern, 17
Hot spot, 60
Hot Springs, 181, 35-36, 154, 175, 177-178
Hot Springs Mammoth Site, 181, 175
Hudson Bay, 181, 13, 72, 74, 122-123
Human kill sites, 98
Human origins, 103-104
Human population growth, 102, 134-135, 145, 161
Humans of Europe, 107, 110
Hunter gatherers, 114
Hunting culture, 110
Hwang Ho River Valley, 181, 117
Hydrocarbon reservoirs, 60
Hydroelectric, 146, 153, 158
Hydrogen sulfide, 150

I

Ice Age, 1, 3-7, 11, 14, 20-21, 23-27, 29-30, 33, 39, 41, 54, 67, 89, 97, 100-102, 109, 111-112, 115, 117-123, 125, 127, 134, 143, 156, 164, 172-174, 176-177
Ice core data, 111
Ice dams, 74
Idealized glacial cycle, 31, 41-42
Identified hearths, 106
Impact of meteorites, 34
Increase crop yields, 158
Increasing atmospheric carbon dioxide, 165
Indirect glacial effects, 75
Insufficient carbon dioxide, 29
Insufficient food, 29
Intensity of twilight, 70
Interglacial periods, 5-6, 32, 38, 41-44, 49-51, 57, 68, 77, 79, 103-105, 108, 110, 112, 119, 143, 156, 170

Internal energy sources, 36
Invertebrate Fossils, 177
Irish elk, 26
Irreversible glacial breakup, 68
Irrigation systems, 115-117
Isostatic readjustment, 74, 136

K

Kansas dust storm, 29
Kansas wheat crop, 128
Kansas Wheat Sketches, 176
Karst Topography, 52
Kilauea gases, 61
Kilauea Volcano, 181, 60-61, 156

L

Land bridge, 11-12, 15, 18-19, 56, 104, 106-107
Land changes, 7, 71-75, 78-79, 114
Large animal genera, 3-4, 7, 24, 26, 30, 98-99, 101, 108-109, 113, 125, 173
Large browsing animals, 23
Large animal extinction, 1, 4, 6, 8-9, 11, 18-19, 21-24, 47, 57, 71, 87-89, 91, 94-95, 97-99, 101, 107, 113, 174
Large grazing animals, 23, 98
Large Pleistocene animals, 18
Last Extinction, xi-xii, 1, 3-5, 9, 17, 19-20, 30-31, 59, 95, 97-98
Last Glacial Breakup, 39, 66, 73, 75, 87-88, 96, 108
Last Glacial Cycle, 5, 11-12, 27, 31, 41, 47, 55, 71-72, 75, 77-79, 82-83, 86, 94, 99, 107, 171-172
Lateral moraines, 74
Lava deposition, 52
Lesser-developed countries, 145-146, 157
Levels of the oceans, 65
Lightning, 36, 62-63, 135
Limestone deposition, 52
Limit of most vegetation, 66, 85-87, 96
Limit of sea ice, 127
Linear volcanoes, 36, 52-53
Liquid natural gas, 151
Little Ice Age, 33, 39, 67, 102, 118-123, 125, 127, 134, 143, 156, 176-177
Load of sediment, 136

Loess Hills, 77, 178
Long-legged North American llama, 23
Long-nosed peccary, 24
Low atmospheric carbon dioxide, 9, 21-23, 26-27, 29, 45, 67, 76, 83, 85-86, 88, 95, 98-99, 101, 108-109, 164

M

Magnetic minerals, 35
Magnetic poles, 35
Maidenhair tree, 83
Mainland China, 12
Mainland of Asia, 120
Mammoth carcasses, 18
Mammoth Epoch, 178
Mammoth steppe environment, 25
Mammoth tooth, 3, 17
Mantle of the earth, 60
Margin of continental glaciers, 39
Marginal plants, 85
Margins of the ocean, 162
Marine Geology, 181, 176
Marsupial lion, 27
Mass extinction, 7, 94-95, 97, 99, 109
Massive carbon dioxide transfer, 9
Mastodon tusks, 19
Mauna Loa Volcano, 181, 156, 161
Maunder minimum, 181, 33, 35, 39, 121
Mears Glacier, 181, 119
Measure absolute carbon dioxide, 65
Mechanics of hurricane generation, 162
Medial moraine, 74
Medieval Warm Period, 181, 118
Melting of continental glaciers, 56
Melting of the glaciers, 4, 6, 31, 57, 67, 79
Mesozoic era, 83, 93
Methane Clathrate, 151, 177
Mid Atlantic Ridge, 181, 36
Mid-latitude glaciers, 53
Middle Eastern Oil Reserves, 181, 152
Migration of plants, 79
Migration routes, 107
Mile thick glacier, 81
Mineral exploitation business, 140
Mississippi Delta today, 79
Mississippi River Valley, 181, 79
Mitochondrial DNA story, 109
Modern beaver, 26
Modern elephant, 17, 23

Moore-Hooper Ellenburger, 181, 60
Moore-Hooper field, 60, 151
Moraine Lake, 181, 72
Most efficient skeletal structure, 104
Most important energy sources, 34
Mound of seashell debris, 112
Mount Assinibone, 72
Mount Kilimanjaro, 181, 39
Mountain glaciers, 5, 37-39, 43, 74, 106, 119
Mountain goats, 24
Mountain sheep, 24
Mountains of southern Chile, 120
Movement of carbon dioxide, 51
Movement of glaciers, 71
Movement of oxygen, 50
Much discussion of euthanasia, 168

N

Naked seed, 83
NASA geologist, 163
Natural disasters, 160
Natural fractures, 60
Natural resources, 144
Nearby ocean source, 38
Net effects, 173
New continental glaciers, 49
New Orleans, 182, 78-79, 85
New source of water, 158
New species, 88, 92-93
New vegetation patterns, 17
New world sabertooth, 24
New Zealand, 13, 102, 120, 154
Next accumulation phase, 172
Next breakup phase, 172
Next glacial cycle, 47, 172
Nile Delta, 182, 117
Nile irrigation, 116
Nile River Valley, 182, 117
Nile Valley irrigation, 116
Non-renewable resources, 144
Non-representative data, 162
Normalized Antarctica data, 162
North American animals, 19, 23
North American camel, 3, 23
North American cheetah, 24
North American elk, 24
North American glyptodont, 24-25
North American horse, 3
North American pampathere, 24-25

Index

North American vegetation, 17
North American wolf, 93
Nuclear fission, 166
Nuclear fusion, 153, 166, 169
Nuclear power, 146, 153, 169
Nuclear power construction, 153
Number of glacial cycles, 5
Numerous land bridges, 11

O

Ocean–atmosphere system, 156
Ocean cores, 41
Ocean level, 11-12, 54, 79
Ocean surfaces, 21-22, 43, 45-47, 53
Oceans rise, 7, 46-47
Ogallala Formation, 182, 131
Oil extraction plant, 148
Oil production industry, 149
Oil refineries, 158
Oil shale, 182, xiv, 149, 165-166, 177-178
Oil shale resource, 149
One-hundred-year climate model forecasts, 163
Orbital plane, 34
Order proboscidea, 17
Origin of Loess, 182, xii, 76
Origin of the earth, 91
Origin of the sediment, 136
Other Times of Extinction, xi, 4
Out of fossil fuel, 168
Outer Continental Shelf, 182, 147
Oxygen concentration, 66, 100

P

Pampean gomphothere, 19, 110
Parallel evolution, 93-94
Particulate matter pollution, 69
Passive solar, 155, 166, 169
Past personal experiences, 1
Peace River, 182, 149
Peak atmospheric carbon dioxide, 170
Peak glacial distribution, 13
Peak of fossil fuel, 168
Peak population, 160, 165, 167-168
Peaking unit, 155
Pennsylvanian Atoka Formation, 152
Periglacial conditions, 13, 39, 84

Period of extinction, 103
Phase of water, 32, 36-38, 173
Photovoltaic cells, 156
Physical properties of water, 1
Plain of Mesopotamia, 116
Plant evolution, 82, 87
Plant leaf area, 87
Plant migration, 11
Pleistocene Extinctions, 182, 177
Pleistocene glacial cycles, 104
Pleistocene vegetational pattern, 14
Polar glaciation of Antarctica, 12
Polar melting, 21-22
Polar-orbit satellites, 162
Polarity of an electromagnet, 35
Population explosion, 7, 108, 111, 125, 174
Population growth, xiii, 128, 134-135, 144-145, 168
Population of plants, 57, 165, 172
Population of the herd, 139
Population of the world, 128, 143, 159, 161
Porous rock, 150
Power generation, 127, 152-155
Prairies of today, 87
Prehistoric Alaska, 182, 178
Previous extinction, 4
Previous glaciation, 5
Prince William Sound, 180, 182, 119, 176
Produce natural gas, 151
Producing surplus food, 122
Production of ethanol, 167
Proved crude oil, 149
Prudhoe Bay, 182, 139

Q

Quaternary Extinctions, 182, 14, 16, 97, 177

R

Radiocarbon dating, 108
Radius of the earth, 65
Rain forest plants, 89
Rapid expansion of population, 8
Rapid positive mental evolution, 103
Raw barometric pressure, 66
Receding glaciers, 39, 114
Record of sunspot activity, 121
Recoverable resources, 150

Recrystallization of snow, 38
Reduction of yields, 171
Reserve life, 165-166
Reward only successful efforts, 148
Richter scale, 119
Rising Sea Level, xii, 79
Rocks of the glaciers, 75
Rotation of the earth, 9, 63-65, 100

S

Saint Lawrence Seaway, 182, 74
San Louis Valley, 182, 77
Sand Dune Deposits, ix, xi-xii, 15, 17, 27, 29-30, 75-78, 87, 96-97, 101, 108
Sand Storm Effects, 182, 29
Sea level drops, 78
Sea level rose, 75, 88
Seashores Sinking, 182, xiii, 136
Seasonal terminology, 32, 41
Secretions of limestone, 62
Sedge plants, 14
Sediment dams, 74, 137
Sequence of events, 21, 75
Shale resources, 149
Shale yields, 149
Shape of the atmosphere, xii, 63, 65
Shape of the skulls, 105
Short-faced kangaroo, 27
Shrub ox, 24
Side-looking radar, 120
Silt buildup, 116
Silt sized particles, 76
Sinking shorelines, 136, 138
Small lung capacity, 66
Small nostrils, 66
Smarter genes, 103
Smothered tundra, 87
Solar energy, 155-156, 166
Solar radiation, 109
Solubilities of atmospheric gases, 50
Solubility of carbon dioxide, 22
Sophisticated mental processes, 104
Source of energy, 35, 155, 169
Source of natural gas, 151
Source of the oxygen, 62
South American bush dog, 25
South American glyptodont, 25
South American nothrothere, 25
South American pampathere, 25

Spectacled bear, 25
Spreading zones, 36, 49, 53
Steam injection, 149
Steppe bison, 26
Steppe environment, 14, 25
Steppes of Eurasia, 26
Stock's pronghorn, 23
Storing water, 116
Stress of glacial cycles, 107
Strongest magnetic field, 35
Studies of air inclusions, 66
Studies of continental drift, 19
Studies of Geysers, 182, 177
Study of carbon dioxide, 8
Study of glaciers, 120
Study of mammoth, 97
Successful strategies, 122
Sufficient carbon dioxide, 8, 65-66, 109, 173
Sufficient oxygen, 63, 92
Sufficient water, 39, 173
Sun side, 66
Sunspot activity, 33, 35, 39, 121
Surfaces of the oceans, 6, 9, 22, 47, 51, 53, 56, 67-68, 96

T

Tar sand, 148-149
Tar sands of Canada, 83
Tasmanian wolf, 93-94
Temperate rain forests, 7
Terminal moraines, 74, 81
Thermal recovery, 149
Thickness of the atmosphere, 63, 92
Three astronomical variations, 34
Tidewater glaciers, 38-39, 43, 119
Times of drought, 44
Times of extinction, xi, 4, 92
Today's vegetation distribution, 17
Top thirty-three feet, 75
Tree lines, 182, xii, 51, 65-67, 76, 85-86, 96, 100
Tundra environments, 94
Twenty-five young orphan elephants, 17
Twenty-foot-long snake, 27

U

U-shaped valleys, 72

Underground aquifers, 158, 161
Underwater volcanoes, 52
Unique isolated population, 105
United States fossil fuels, 165
Unloading of the ice, 137
Uranium ore, 109
Use coal, 166, 169
Use natural gas, 151
Use of solar energy, 155
Use of wind, 155
Use passive solar energy, 166
Utility power grids, 155

V

Varieties of food plants, 128
Vegetation pattern, 125
Very large animal extinction, 98
Viable concept, 55
Void spaces, 37

W

Wastage of the glaciers, 46, 53
Waste-to-energy plants, 154
Water, 134, 158
Water buffalo, 26
Water heating relationship, 38
Water resources, 183, 132, 178
Weathering of the rocks, 61
Western hemlock, 88
Western North American camel, 23
Wheat crop sketches, xiii, 129
Wheat yield, 131
Widespread seafaring skills, 107
Woodland muskox, 23
World food-productive capacity, 132
World's fifteen largest-capacity reservoirs, 138

Y

Yearly heat cycles, 31